Wireshark 2 Quick Start Guide

Secure your network through protocol analysis

Charit Mishra

BIRMINGHAM - MUMBAI

Wireshark 2 Quick Start Guide

Commissioning Editor: Vijin Boricha
Acquisition Editor: Reshma Raman
Content Development Editor: Aditi Gour
Technical Editor: Shweta Jadhav
Copy Editor: Safis Editing
Project Coordinator: Hardik Bhinde
Proofreader: Safis Editing
Indexer: Aishwarya Gangawane
Graphics: Jason Monteiro
Production Coordinator: Deepika Naik

First published: June 2018

Production reference: 1200618

Published by Packt Publishing Ltd.
Livery Place
35 Livery Street
Birmingham
B3 2PB, UK.

ISBN 978-1-78934-278-9

www.packtpub.com

`mapt.io`

Mapt is an online digital library that gives you full access to over 5,000 books and videos, as well as industry leading tools to help you plan your personal development and advance your career. For more information, please visit our website.

Why subscribe?

- Spend less time learning and more time coding with practical eBooks and Videos from over 4,000 industry professionals

- Improve your learning with Skill Plans built especially for you

- Get a free eBook or video every month

- Mapt is fully searchable

- Copy and paste, print, and bookmark content

PacktPub.com

Did you know that Packt offers eBook versions of every book published, with PDF and ePub files available? You can upgrade to the eBook version at `www.PacktPub.com` and as a print book customer, you are entitled to a discount on the eBook copy. Get in touch with us at `service@packtpub.com` for more details.

At `www.PacktPub.com`, you can also read a collection of free technical articles, sign up for a range of free newsletters, and receive exclusive discounts and offers on Packt books and eBooks.

Contributors

About the author

Charit Mishra is an ICS/SCADA professional, working as a security architect for critical infrastructure across several industries, including oil and gas, mining, utilities, renewable energy, transportation, and telecom. He has been involved in leading and executing complex projects involving the extensive application of security standards, frameworks, and technologies. A postgraduate in computer science, Charit's profile boasts of leading industry certifications such as OSCP, CEH, CompTIA Security+, and CCNA R&S. Moreover, he regularly delivers professional training and knowledge sessions on critical infrastructure security internationally.

About the reviewer

Anish has a YouTube channel named *Zariga Tongy* where he loves to post videos on security, hacking and other cloud related technology.

Packt is searching for authors like you

If you're interested in becoming an author for Packt, please visit `authors.packtpub.com` and apply today. We have worked with thousands of developers and tech professionals, just like you, to help them share their insight with the global tech community. You can make a general application, apply for a specific hot topic that we are recruiting an author for, or submit your own idea.

Table of Contents

Preface

Wireshark is the world's most popular free and open source protocol analyzer, and it is commonly used by networking and security professionals for troubleshooting, analysis, protocol development, and forensics. The primary objective of Wireshark is to capture network traffic and display the packet data in, as detailed a way as possible. It helps professionals view the content of network traffic on a microscopic level.

This book is written from the standpoint of using Wireshark and learning how network protocols function and provides a practical approach to conducting protocol analysis, troubleshooting network anomalies, and examining security issues. I have tried to depict common scenarios that you may come across in day-to-day operations through practical demonstration wherever possible to help you understand the concepts better. By reading this book, you will learn how to install Wireshark, work with Wireshark GUI elements, and learn some advanced features behind the scenes, such as the filtering options, the statistics menu, and decrypting wireless and encrypting traffic. You can be the superhero of your team who helps resolve connectivity issues, network administration tasks, and computer forensics because *Packets Are Life*. If your routine job requires dealing with computer networks and security, then this book will give you a strong head start. Happy sniffing!

Who this book is for

This book is for students/professionals who have basic experience and knowledge of the networking and who want to get up to speed with Wireshark in no time. This book will take you from the installation to the usage of commonly used tools/tricks. The book will get you comfortable with the GUI elements of Wireshark and explain the fundamentals of the science behind protocol analysis.

What this book covers

Chapter 1, *Installing Wireshark,* will provide you with an introduction to the basics of the TCP/IP model and a step-by-step walk-through of the installation of Wireshark on your favorite operating system.

Chapter 2, *Introduction to Wireshark and Packet Analysis*, will help you understand the basics and science behind packet analysis, as Wireshark come in handy and proves to be a Swiss Army knife for professionals dealing with network, security, and digital forensics. In this chapter, you will also understand the trick of placing the sniffer in a strategic location to get most out of your network.

Chapter 3, *Filtering Our Way in Wireshark*, will help you identify and apply the Wireshark filters, namely the capturing and displaying filters. Filtering provides a powerful way to capture or see the traffic you desire; it's an effective way to remove the noise from the stream of packets we desire to analyze.

Chapter 4, *Analyzing Application Layer Protocols*, will help you understand the approach and methodology for analyzing application layer protocols such as HTTP, SMTP, FTP, and DNS through Wireshark. As we know, application layer protocols typically interface between a client and a server. It is critical to understand the structure and behavior of application layer protocols packets in order to identify anomalies with efficiency.

Chapter 5, *Analyzing the Transport Layer Protocols TCP/UDP*, will help you understand the underlying network technology, enabling the movement of network packets across routing infrastructures through the analysis of transport layer protocols such as TCP and UDP. TCP and UDP are the basis of networking protocol, and it is important to understand their structure and behavior.

Chapter 6, *Network Security Packet Analysis*, will guide you through using Wireshark to analyze security issues, such as analyzing malware traffic and footprinting attempts in your network.

Chapter 7, *Analyzing Traffic in Thin Air*, will help you in understand the methodology and approach involved in performing wireless packet analysis. This chapter shows you how to analyze wireless traffic and pinpoint any problems that may follow. We will also learn the cool trick of decrypting wireless traffic using Wireshark.

Chapter 8, *Mastering the Advanced Features of Wireshark*, will provide you with insight into the advanced options and elements available in Wireshark, such as a statistics menu, and will also provide a brief and summarized approach on how to work with command-line packet sniffing applications, such as Tshark.

To get the most out of this book

- Basic understanding of networking protocols, OSI and TCP/IP model
- A computer system with a basic internet connection to follow the depicted scenarios

Download the color images

We also provide a PDF file that has color images of the screenshots/diagrams used in this book. You can download it here: https://www.packtpub.com/sites/default/files/downloads/Wireshark2QuickStartGuide_ColorImages.pdf.

Conventions used

There are a number of text conventions used throughout this book.

CodeInText: Indicates code words in text, database table names, folder names, filenames, file extensions, pathnames, dummy URLs, user input, and Twitter handles. Here is an example: "Mount the downloaded WebStorm-10*.dmg disk image file as another disk in your system."

Bold: Indicates a new term, an important word, or words that you see onscreen. For example, words in menus or dialog boxes appear in the text like this. Here is an example: "Select **System info** from the **Administration** panel."

 Warnings or important notes appear like this.

 Tips and tricks appear like this.

Get in touch

Feedback from our readers is always welcome.

General feedback: Email feedback@packtpub.com and mention the book title in the subject of your message. If you have questions about any aspect of this book, please email us at questions@packtpub.com.

Errata: Although we have taken every care to ensure the accuracy of our content, mistakes do happen. If you have found a mistake in this book, we would be grateful if you would report this to us. Please visit www.packtpub.com/submit-errata, selecting your book, clicking on the Errata Submission Form link, and entering the details.

Piracy: If you come across any illegal copies of our works in any form on the Internet, we would be grateful if you would provide us with the location address or website name. Please contact us at copyright@packtpub.com with a link to the material.

If you are interested in becoming an author: If there is a topic that you have expertise in and you are interested in either writing or contributing to a book, please visit authors.packtpub.com.

Reviews

Please leave a review. Once you have read and used this book, why not leave a review on the site that you purchased it from? Potential readers can then see and use your unbiased opinion to make purchase decisions, we at Packt can understand what you think about our products, and our authors can see your feedback on their book. Thank you!

For more information about Packt, please visit packtpub.com.

Installing Wireshark

1

This chapter provides you with an introduction to the basics of the TCP/IP model and a step-by-step walkthrough of how to install Wireshark on your favorite operating system. You will be introduced to the following topics:

- What is Wireshark?
- A brief overview of the TCP/IP model
- Installing and running Wireshark on different platforms
- Troubleshooting common installation errors

Introduction to Wireshark

Wireshark is an advanced network and protocol analyser, it lets you visualize network's activity in graphical form, and assists professionals in debugging network-level issues. Wireshark enhances the ability of network and security professionals by providing detailed insight into the network traffic. However, Wireshark is also used by malicious users to sniff network traffic in order to obtain sensitive data in the form of plain text.

Why use Wireshark?

Many people, including myself, are obsessed with the simplicity of the packet-capturing features that Wireshark provides us with. Let's quickly go through a few of the reasons why most professionals prefer Wireshark to other packet sniffers:

- **User friendly**: The interface of Wireshark is easy to use and understand, tools & features are very well organized and represented.
- **Robustness**: Wireshark is capable of handling enormous volumes of network traffic with ease.

- **Platform independent**: Wireshark is available for different flavors of operating system, whether Windows, Linux, and Macintosh.
- **Filters**: There are two kinds of filtering options available in Wireshark:
    - You choose what to capture (**capture filters**)
    - You choose what to display after you've captured (**display filters**)
- **Cost**: Wireshark is a free and open source packet analyzer that is developed and maintained by a dedicated community of professionals. Wireshark also offers a few paid professional applications as well. For more details, refer to Wireshark's official website `https://www.wireshark.org/`.
- **Support**: Wireshark is being continuously developed by a group of contributors that are scattered around the globe. We can sign up to Wireshark's mailing list or we can get help from the online documentation, which can be accessed through the GUI itself. Various other online forums are also available for you to get the most effective help; go to Google Paid Wireshark Support to learn more about the available support.

The installation process

The installation of Wireshark is very simple and easy to follow. Go through the following steps to install it on your system:

1. The recipes and examples in this book will be for use on a Macintosh and Windows PC; for other operating systems, the installation is the same. Some OSes, such as Kali Linux, come with a preinstalled version of Wireshark.
2. Once you have located the correct version of Wireshark for your platform (Wireshark 2.6.1 Intel 64.dmg), install Wireshark by following the wizard.
3. Restart the computer after completion of the installation process to commit the changes that were made.
4. Double-click the Wireshark icon on your desktop to the run the application:

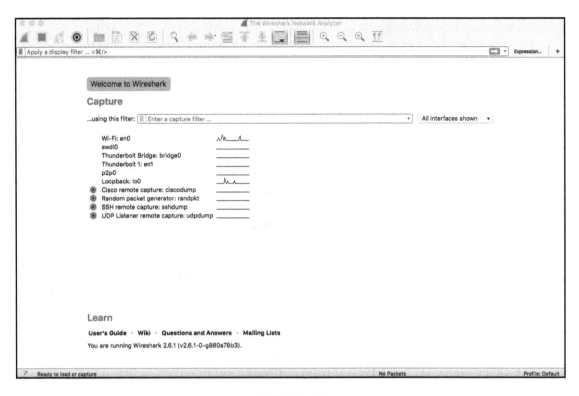

The Wireshark screen

Troubleshooting common installation errors

Go through the following simple checklist to ensure that you are able to run Wireshark successfully (make sure that all of these criterias are met):

- You have downloaded Wireshark from known and trusted source only
- You have administrative privileges to run Wireshark
- The installation of Wireshark and the Winpcap driver has been completed successfully without any exceptions
- You are connected to the network that you want to capture network traffic from
- If you are trying to sniff using a virtual machine, ensure that you have set your network adapter to bridged mode
- Restart your machine to ensure the changes have been applied after successful installation of Wireshark

- Your NIC card supports promiscuous mode sniffing (when needed)
- You can see all of the interfaces (wired, wireless, and logical) on the home screen of Wireshark
- The line graph followed by the interface name shows activity on the Homescreen
- Also, you have legal permissions to capture network traffic

A brief overview of the TCP/IP model

The world of network communication is governed by a set of protocols (rules and regulations) in order to function as intended. Protocols govern the transmission of network packets/segments/frames over a communication channel between endpoints. In order to understand how network packets stick together, forming a stream of traffic, we need to understand the basics of the networking that is the TCP/IP model. The TCP/IP model was originally known as the DoD model, a project that was regulated by the United States Department of Defense. All of the communication that we witness over the internet and other networks happens only through TCP/IP.

The TCP/IP model takes care of every part of packet's life cycle, namely, how a packet comes to life, how a packet is generated, how information pertaining to packet gets attached data payload (PDU), how it is routed through intermediary nodes, linking with other packets and so on.

It is strongly recommended to do some self-study on TCP/IP and how it functions, before you proceed ahead, as this book requires decent amount of familiarity with protocols.

The layers in the TCP/IP model

The TCP/IP model comprises four layers, as shown in the following diagram. Each layer has a specific purpose to fulfill and utilizes a set of protocols to facilitate communications. Every protocol in every layer has a specific purpose:

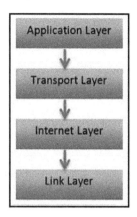

The first layer is the **Application Layer**, which directly interacts with users and subsequent layers and protocols; it is primarily concerned with the representation of the data in a understandable format to the user. The application layer also keeps track of user sessions, monitoring who is connected; it uses a set of protocols that helps to interface with users and other layers in the TCP/IP model. Some popular protocols in the Application Layer are as follows:

- **Hypertext Transfer Protocol (HTTP)**
- **File Transfer Protocol (FTP)**
- **Simple Network Management Protocol (SNMP)**
- **Simple Mail Transfer Protocol (SMTP)**

The second layer is the **Transport Layer**. The purpose of this layer is to create sockets (a combination of the port and IP address) in order to let two endpoints communicate. Sockets facilitate the creation of multiple distinct connections between two or more devices (more than one tab can be opened in Chrome).

An IP address is required for communication between devices in different networks/segments (such as is used between two router interfaces or communication over the internet). It can also be used in **local area network (LAN)** communication, and is established over physical addresses (MAC). Apart from the restricted range of port numbers, operating systems and applications can choose a random port (other than ports 1 to 1013) for communication.

The transport layer also serves as a backbone for the communication. The two most critical protocols that work in this layer are the TCP and UDP:

- The **TCP** is a connection-oriented protocol, also called a reliable protocol. Firstly, a dedicated communication channel is established between the endpoints, which is then followed by data transmission. Equally partitioned chunks are transmitted from the source, and the receiving end sends an acknowledgement for every packet received. The side that is sending the data resends the packet if an acknowledgement is not received within a stated time frame.
- The **UDP** is a connectionless protocol and is often called an unreliable communication form. In the UDP, no dedicated channel is established, which also makes it a simpler and faster way of communication. There are also no acknowledgement packets sent by the endpoints. For example, if you are playing an online game, the loss of a few packets over the communication channel is not going to hamper your gaming experience because the number of packets coming through is huge, and a few missing packets will not make much difference to the overall quality of the network stream.

The third layer is the **Internet Layer**, which is primarily concerned with routing and movement of data between networks. The primary protocol that works in this layer is the **IP (Internet Protocol)**. The IP provides the network packets with the routing capability that they need in order to reach their destination. Other protocols included in this layer are the ICMP and IGMP.

The fourth and final layer is the **Link Layer** (often called the network interface layer). It interfaces with the physical network hardware. There are no protocols specified in this layer by the TCP/IP; however, several protocols are implemented, such as the **Address Resolution Protocol** (ARP) and the **Point to Point Protocol**(PPP). This layer is concerned with how information travels inside the communication channel (wired or wireless). The link layer is responsible for establishing and terminating the connection, as well as converting the signals from analog to digital and vice versa. Devices such as bridges and switches operate in this layer.

As data progresses from the application layer to the link layer, several bits of information are attached to the data in the form of headers or footers, which allow different layers of the TCP/IP to communicate with each other. The process of adding these extra bits is called data encapsulation, and in this process, a **protocol data unit (PDU)** is created at the end of the networking process (passing through the application to the link layer).

PDU consists of the data along with network addressing and protocol information that gets attached as part of the header or footer. By the time PDU reaches the bottom-most layer, it is embedded with all the required information necessary for transmission. Once the PDU reaches the destination, the attached header and footer PDU elements are ripped off one by one as it passes through each layer of the TCP/IP model and progresses upward in the model.

The following diagram depicts the process of encapsulation:

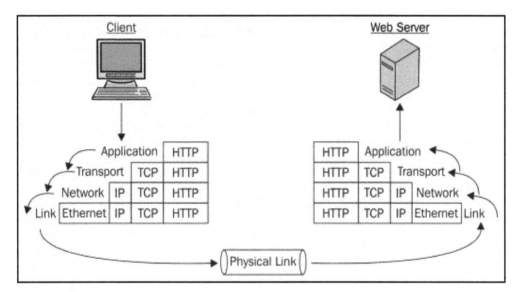

Summary

In this chapter, we looked at the basic networking concepts that you need to know, along with an introduction to Wireshark. Wireshark is a protocol analyzer that is used worldwide by technology professionals to capture and analyze network-level packets.

We also learned about the TCP/IP model. The TCP/IP model has four layers: the application layer, transport layer, network layer, and the link layer. Data is encapsulated as it passes from one layer to another; the resulting packet at the bottom is called a complete PDU.

The TCP is a reliable protocol because acknowledgements are sent as part of its process, whereas the UDP is an unreliable protocol because no acknowledgements are sent.

To install Wireshark, you just need to visit `http://www.wireshark.org` and then download the appropriate version for your operating system.

Troubleshooting your Wireshark can be done by ensuring that the network is working fine, that you have the full rights required to install and run the application, and that the installation had completed without any exceptions.

In the next chapter we will run our first Wireshark capture and get to feel the protocol analysis experience.

Introduction to Wireshark and Packet Analysis

2

This chapter will help you to understand the basics and science behind packet analysis. Wireshark comes in very handy and proves something of a Swiss knife for professionals dealing with network, security, and digital forensic roles. You will learn about the following topics in this chapter:

- Introduction to Wireshark
- How Wireshark works
- Capturing methodologies
- Understanding the GUI of Wireshark
- Starting our first capture

What is Wireshark?

Wireshark is a packet-sniffing application that is used by IT professionals for a diverse set of requirements (including forensics, troubleshooting, and enhancing network performance). You can download it for free from `https://www.wireshark.org/download.html`, where it is available for the majority of platforms, including Linux, Macintosh, and Windows.

Packet sniffing is also referred to as tapping into the wire, which basically involves reading pieces of information traveling in a communication channel. Considerations such as placement of sniffer, protocols to be analyzed, and communication channel type need to be assessed before capturing network packets.

How Wireshark works

Wireshark collects network traffic from the wire through the computer's network interface, running in promiscuous mode (if needed), to inspect and display information related to protocols, IP addresses, ports, headers, and packet length. The following diagram is an illustration of how all the elements work together to display packet-level information to the user (source: `https://www.wireshark.org`):

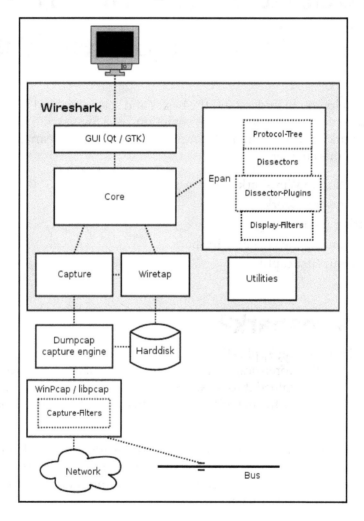

Wireshark comes with the **Winpcap/libcap** driver, which enables NIC to the run in promiscuous mode; the only time you don't have to sniff in promiscuous mode is when the packets are directly, intentionally destined/generated to and/or from your device.

On operating systems, you should have privileges to run Wireshark. There are three processes that every protocol analyzer follows: collect, convert, and analyze. These are described as follows:

- **Collect**: Choose an interface to listen to traffic and capture network packets.
- **Convert**: Increase the readability of non-human-readable data. Packets are converted to easily understood information through a GUI.
- **Analyze**: Analyze network traffic pertaining to the packets, protocols, raw data and more through the usage of statistical and graphical features.

As discussed in the previous chapter, protocols are the set of rules and regulations that govern the process of communication between two network devices and control the environment under which they operate.

An introduction to packet analysis with Wireshark

Packet/traffic analysis deals with the study of network traffic, where the objective is to understand the structure, movement, and behavior of packets. Packet analysis is performed over live traffic or done over an already captured stream of traffic.

Numerous issues arise in day-to-day networking infrastructures, and if you are responsible for handling the network or security of your digital environment, you need to equip yourself with troubleshooting and analytical tools. Most of the issues escalate and are rectified at the packet level in networking. Issues arising at the packet level can gradually end up disrupting critical business communication, leading to loss of revenue. Even the best networking hardware utilizing the most advanced and secure set of protocols and services can go against you or behave abnormally. To perform a root cause analysis in such situations, you might need to dig down to the packet level in order to understand the anomaly. Packet analysis can be used for the following purposes:

- To analyze network issues by looking into the packets and their headers to gain better insights.

- To detect and analyze network intrusion attempts through filtering patterns and signatures.
- To detect network misuse by internal or external users by establishing firewall rules in your security appliance and then monitoring those rules.
- To study and isolate exploited systems so that the affected system doesn't become a pivot point.
- To monitor and analyze data in motion as it travels live in the wires of your network.
- To have better control over the allowed and restricted categories of information traveling in your network. For instance, say you want to create a rule in the firewall that will block access to torrent sites (peer-to-peer file sharing). Blocking access to them can be done from your manageable router through access lists also, but the origin of such packets can be identified and validated through traffic analysis.
- To gather and report network statistics by filtering packet trails.
- To learn who is on a live network and what they are doing (they may be consuming network bandwidth or trying to connect to restricted websites), and to learn whether someone is trying to bypass the network restrictions you configured.
- To debug client/server communications so that all the requests and replies communicated on your network can be audited.
- To identify applications that are sitting in the corner of your network and consuming the bandwidth. They might be making your network insecure, unresponsive, or visible to the public network.
- To debug network protocol implementations and any anomalies being generated due to unintentional misconfigurations errors or human error.
- To identify abnormal/malicious traffic patterns that your network, then to analyze, control/supervise, and make yourself ready for such events.

When performing packet analysis, the things to be considered are as follows:

- The protocol(s) to be interpreted
- Whether you need to capture traffic from all sources and all destinations
- Placing your sniffer adequately
- Capturing traffic pertaining to a particular port or service to avoid unwanted noise

You should record and build use cases pertaining to the network traffic pattern and behavior. Use cases may assist engineers in troubleshooting network issues.

Packet analyzers can interpret most networking protocols (such as IP and ICMP), transport-layer protocols (such as TCP and UDP), and application-layer protocols (such as DNS and HTTP).

How to do packet analysis

Network packets are captured in raw binary form, and passed through the wiretap library and capture engine, and then to the core engine, with its dissector plugins and filters. The translated data is then displayed in packet frames through **Graphical Toolkit** (**GTK**).

Capturing methodologies

In order to capture the right set of a packets stream, you would need to know where to place your protocol analyser. Depending on the requirements (source of packets, number of packets, type of packets, and more), a protocol analyzer needs to be placed at a certain point in the network. Also, a few configuration changes in a network device may be necessary, such as switch configuration changes (mirroring is done in network switches to capture packets from one or more sources). The following sub sections discuss a few means of assessing the best way of configuring protocol analyses in certain types of topology.

Hub-based networks

It is relatively easy to sniff in a hub-based network topology, because you've got the freedom to place the sniffer at any place you want, as hubs are designed to broadcast each and every packet to all connected devices.

However, due to such design deficiencies, hub-based network topologies face issues in terms of overall performance. Network hubs do not have much capability in terms of prioritizing or forwarding traffic to specific ports only. They often become victims of collision-related problems. For instance, if more than one device connected to a hub start sending data at the same time, there is a high a probability that the packets will collide and fail to reach their destination. The sending side will be informed of dropped packets, which will then be re-sent, but it will cost the network and its administrator in time, improper bandwidth utilization, and performance issues.

The switched environment

Due to relatively few restrictions present in switch-based infrastructures, packet analysis becomes quite challenging. Like hubs, switches do not broadcast the packets to every network port except the port the packet is received from. They learn the physical addresses of devices through the **ARP** (**address resolution protocol**) and populate a list of port numbers with corresponding MAC addresses. Even so, through some hardware or configurational changes it is possible to capture packets from other ports. The two most popular techniques are hubbing out and port mirroring.

In order to capture the stream of packets coming from one or more ports, configure port mirroring using the switch configuration console. Most intelligent switches give the option to configure it through an easy-to-understand graphical interface.

Let's make it simpler for you with a logical illustration. For instance, let's assume that we have a 24—port switch and eight PCs, which are connected to different switch ports. We can place our sniffer (Wireshark PC) in any of the free switch ports and then configure port mirroring, which will copy all the traffic from the desired device we want to sniff to the port of our choice. The following screenshot shows the set of commands used in a Cisco Switch to configure port mirroring:

```
Switch(config)#monitor session
Switch(config)#monitor session 1 sou
Switch(config)#monitor session 1 source in
Switch(config)#monitor session 1 source interface fa0/2
Switch(config)#monitor session 1 des
Switch(config)#monitor session 1 destination in
Switch(config)#monitor session 1 destination interface fa0/4
Switch(config)#exit
```

So, let's understand it better: in the previous screenshot, I have configured what to listen to all the packets originating from port `fa0/2` to port `fa0/4`. Port `fa0/2` will be the target machine and port `fa0/4` will be a Wireshark machine.

Once this is completely configured, we will be able to easily sniff and analyze network packets flowing back and forth from port `fa0/2`. This technique is one of the easiest to configure; the only thing you need to know beforehand is how to work with network devices.

The following diagram depicts a simple demonstration of port mirroring:

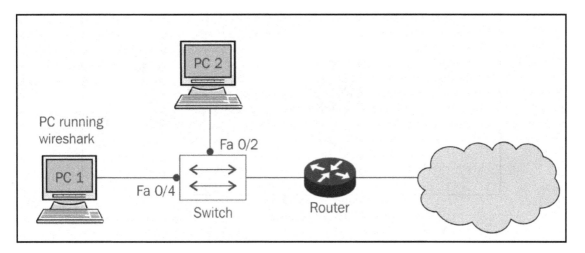

Port mirroring

Hubbing out is feasible when your switch doesn't support port mirroring. To use the technique, you must actually unplug the target PC from the switched network, then plug your hub to the switch, and then connect your analyzer and target device to the hub so the target device becomes part of the same network.

Now the protocol analyzer and the target machine are part of the same broadcast domain. The following diagram will make it easier for us to understand the process precisely and in a simpler way:

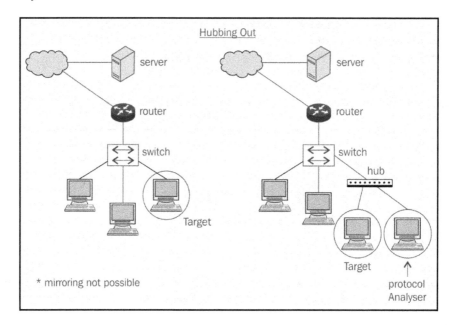

Hubbing out

ARP poisoning

Poisoning the ARP table entries of a device and then forwarding them through your machine is one unethical way of capturing the traffic from the target machine.

Let's say, for example, we have the default gateway at IP 192.168.1.1 and one client machine configured at IP 192.168.1.2. Both of these devices are maintaining local ARP cache entries. That enables them to send packets over the LAN. Now, the Wireshark (use arpspoof or ettercap to poison the ARP entries) machine at IP 192.168.1.3 will poison the ARP cache entries by flooding the client and gateway machine with multiple ARP packets, stating to the client PC that the default gateway has been changed to IP 192.168.1.3 and stating the gateway that the client is now at IP 192.168.1.3; this will make every packet go through the Wireshark machine.

The command to view the ARP cache in your PC/router/server, which will display MAC addresses associated for a particular IP address, is `arp -a`. Have a look at the normal ARP entries:

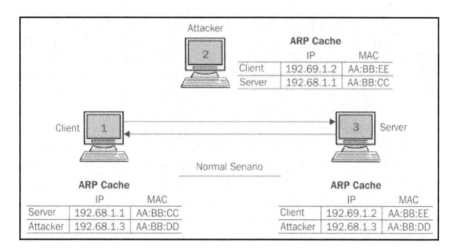

ARP poisoning (the normal scenario)

Here is how the entries will look before the ARP is poisoned:

```
Before ARP is Poisoned

192.68.1.1 - (Server)
192.68.1.2 - AA:BB:EE
192.68.1.3 - AA:BB:DD

192.68.1.2 - (Client)
192.68.1.1 - AA:BB:CC
192.68.1.3 - AA:BB:DD

192.68.1.3 - (Attacker)
192.68.1.1 - AA:BB:CC
192.68.1.2 - AA:BB:EE
```

Now that you've understood what the ARP is and how it works, we can try to poison the **ARP Cache** of both the default gateway and the client with the attacker's MAC address. In simple terms, we will replace the client's MAC address in the default gateway's ARP cache with the attacker's MAC address. We will do the same in the client's MAC address, replacing the default gateway's MAC address with the attacker's MAC address. As a result, every packet destined to the client from the default gateway back and forth will be sent through the attacker's machine. Below are the ARP entries from the client, the server, and the attacking machine after a successful poisoning attack.

```
After ARP is Poisoned

192.68.1.1 - (Server)
192.68.1.2 - AA:BB:DD
192.68.1.3 - AA:BB:DD

192.68.1.2 - (Client)
192.68.1.1 - AA:BB:DD
192.68.1.3 - AA:BB:DD

192.68.1.3 - (Attacker)
192.68.1.1 - AA:BB:CC
192.68.1.2 - AA:BB:EE
```

The poisoned machines will not be able to determine whether their ARP has been modified unless checked proactively. The following diagram depicts the ARP table entries of all three systems involved in the MiTM attack scenario:

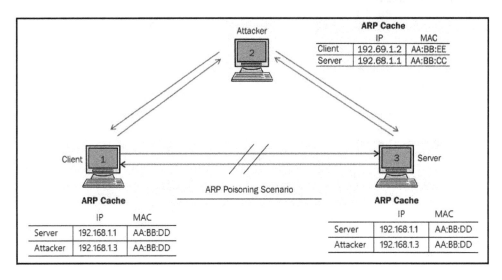

ARP poisoning (the poisoned scenario)

Other than these two techniques, there is a variety of hardware available on the market popularly known as taps, which can be placed between any two devices to sniff and analyze the traffic. Though this technique is effective in capturing network traffic in some scenarios, it should only be practiced or deployed in an authorized and controlled environment, because of its malicious nature.

Passing through routers

When dealing with routed environments, the important aspect of packet analysis would be to place our sniffer at the suitable place from where we can capture the desired traffic packets. Dealing with routed structures demands more skills in terms of networking technologies, and certainly in terms of routers. Consider the following hypothetical routed environment for the sake of understanding.

Router 1, router 2, and router 3 are working together; each of them handles traffic for at least 2-3 PCs. Router 1 is acting as a root node while controlling routing for its child networked nodes (router 2 and router 3).

Router 3 PCs are not able to connect to router 1 PCs. To resolve this issue, the admin places a sniffer (protocol analyzer) inside the router 3 area, and starts analyzing the traffic, but is not able to figure out the anomaly that is causing downtime. The admin decides to change the protocol analyzer location and moves to the router 1 area, and now follows similar steps for troubleshooting. After a while, they figure out what the issue was and troubleshoot it successfully.

The conclusion is that placing the sniffer in your networked infrastructure is quite a critical decision and task.

After reading this, I hope we've a fair amount of knowledge on how protocol analyses are done in certain topologies. Now, let's see what the Wireshark interface looks like, and how we can initiate capturing network packets.

If you do not have Wireshark installed, you can get a free copy from `https://www.wireshark.org/download.html`. To walk through the demonstrations in this book, you also need to be familiar with the interface.

The Wireshark GUI

Before we discuss its awesome features, let's talk about some of critical events in the Wireshark domain.

Wireshark was built during the late 1990s. Gerald Combs, a young college graduate from Kansas City, developed Ethereal (the basic version of Wireshark), and by the time Combs developed this awesome invention, he had landed himself a job. After a few years of service, Combs decided to quit his job and pursue his dreams by developing Ethereal further. Unfortunately, as per the legal terms, Combs' invention was part of the company's proprietary software. Despite this, Combs left the job and started working on the new version of Ethereal, which he titled Wireshark. Since 2006, Wireshark has been in active development and is being used worldwide. It supports more than 800 protocols both corporate IT and **ICS (industrial control system)**.

Before we go ahead and start the first capture, we need to get a bit familiar with the options and menus available.

There are six main parts in the Wireshark GUI, which are explained as follows:

- **Menu Bar**: This represents tools in a generalized form, which are organized in the **Applications** menu.
- **Main Tool Bar**: This consists of the frequently used tools/features that offer efficient utilization of the software.
- **Packet List Pane**: This displays all the packets getting captured by Wireshark.
- **Packet Details Pane**: This is used to view details pertaining to the selected packet from the Packet List Pane. Detailed information regarding the packets is divided into categories corresponding to each layer of the TCP/IP model. This can be used to view source and destination IP addresses and different protocols used for communication arranged in the bottom-to-top approach (link layer to application layer).
- **Bytes Pane**: This shows the data in the packets in the form of hex bytes and their corresponding ASCII values; it shows the values in the form in which they travel in the wires.
- **Status Bar**: This displays details such as total packets captured.

The following screenshot will help you to identify different sections in the application; please make sure that you get yourself acquainted with all of them before proceeding further:

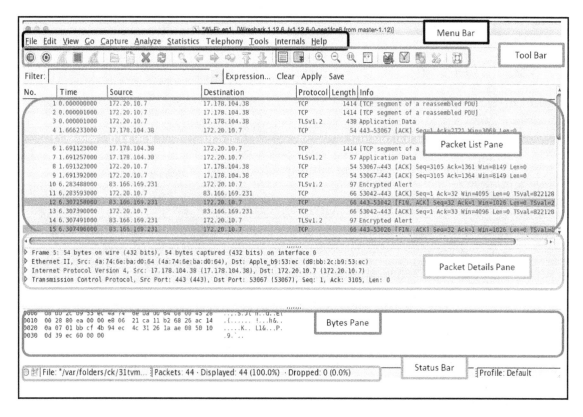

Within the toolbar area, we have a few useful tools. I would like to give you a brief overview of some of them:

- ⊚ : Choose an interface for listening

- ◉ : Customize the capture process (interface, filters, and so on)

- ◢ ▣ ◣ : Start/stop/restart the capturing process

- 🗁 : Open a saved capture file

- 🗋 : Save the current capture in a file

- ⟳ : Reload the current capture file

- ✖ : Close the current capture file

- ⬅ : Go to previous packet

- ➡ : Go to next packet

- ⬅ : Go to a specific packet number

- ▤ : Toggle color coding for the packets on/off

- ▾ : Toggle the auto scroll on/off

- 🔍 🔍 🔍 : Zoom in, zoom out, and reset zoom to the default

- ▦ : Change the color coding as per requirements

- ▼ : Narrow down the window to capture packets

- ▽ : Configure display filters to only see what is required

Even after selecting the interface, there can sometimes not be any packets listed in the list pane; there can be multiple reasons for this, some of which are as follows:

- You do not have any network activity
- Your interface is not able to capture the desired packets, due to privileges
- You do not have promiscuous mode activated or do not have an option for promiscuous mode

Once you click on the **Capture** button in the tool pane, Wireshark will start capturing and you will be able to see some traffic activity colored with different codes, protocol names, packet numbers, IP addresses, and so on:

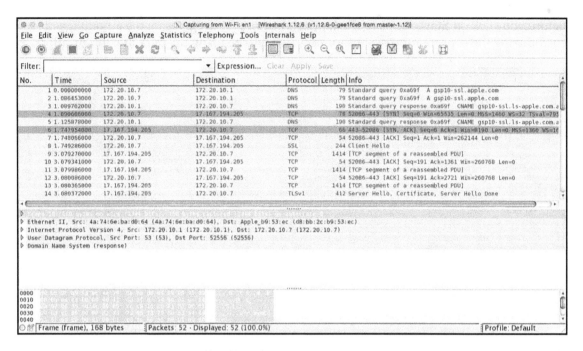

The Wireshark capture screen

Starting our first capture

As you've been introduced to the basics of Wireshark and since you have learned how to install Wireshark, I feel you are ready to initiate your first capture. I will be guiding you through the following series of steps to start/stop/save your first Wireshark capture:

1. Open the Wireshark application.
2. Choose an interface to listen to.

Before you click on **Start**, we have the **Options** button, which gives us the advantage of customizing the capture process; but for now, we will be using the default configuration:

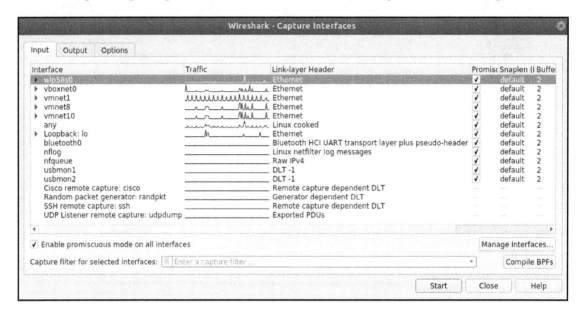

The capture customization screen

Below are the steps for the capture process:

1. Click on the **Start** button to initiate the traffic capture.
2. Open a browser.
3. Visit any website in your browser to generate some HTTP-based traffic:

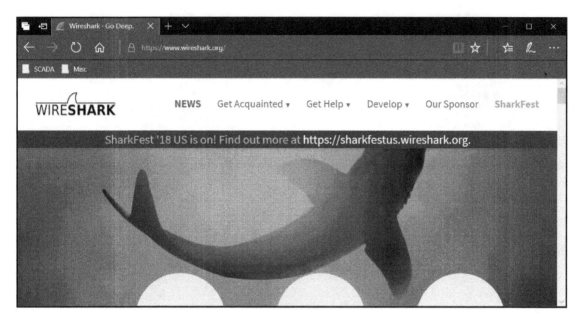

The Wireshark website

4. Switch back to the Wireshark screen; if everything goes well, you should be able to see numerous packets getting captured in your Wireshark GUI inside the Packet List Pane.

5. To stop the capture, you can just click on the **Stop** capture button in the toolbar. area, or you can click on **Stop** under the **Capture** menu bar:

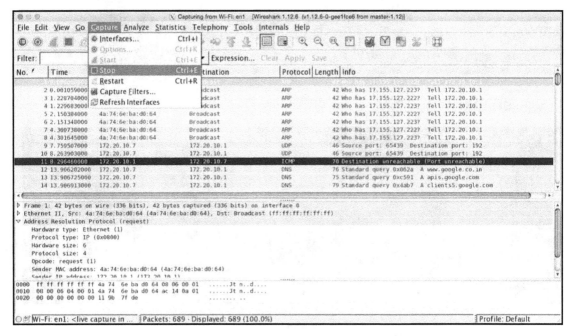

Stopping capture

6. Now, the last step is to save the capture file for later use.

7. Save your file with the default `.pcapng` extension in your folder.

If you have read all the steps all the way up to this point, I would encourage you to create your first capture file and save it in some workspace of your choice.

Summary

This chapter laid the foundation of basic networking concepts and gave an introduction to the Wireshark GUI. Wireshark is a protocol analyzer that is used worldwide by IT professionals to capture and analyze network-level packets.

The Wireshark GUI is user-friendly, robust, and platform-independent; even new IT professionals can easily adopt the tool.

One important aspect of protocol analyzing is to place the sniffer at the right place; every organization's infrastructure is different, so we might need to apply different techniques in order to get the right packets to use.

Hubbing out, port mirroring, ARP poisoning, and tapping are some of those useful techniques that can be used to monitor and analyze traffic in different situations.

There are six main parts in the Wireshark tool window: **Menu Bar**, **Main Tool Bar**, **Packet List Pane**, **Packet Details Pane**, **Bytes Pane**, and **Status Bar**.

Using the back/forward key during a packet analysis scenario can be really useful. You should know about all the tools that are displayed in the main toolbar area.

In the next chapter, you will learn how to work with the different kinds of filters available in Wireshark.

Filtering Our Way in Wireshark

3

This chapter will assist you in identifying and applying the usage of Wireshark filters—namely, the capture and display filters. Filtering provides a powerful way to capture or see traffic; it is an effective way to segregate the desired traffic stream from noise (traffic). The following are the topics we will cover in this chapter:

- Introducing capture filters
- Why and how to use capture filters
- Introducing display filters
- Why and how to use display filters
- Colorizing traffic

Let's start our analyzer and apply some filters to understand the usage and effectiveness of them. We will take a step-by-step walk through the process of creating display and capture filters. Also, we will find utility, which is quite effective when troubleshooting network issues.

Introducing filters

The two types of filters offered by Wireshark are capture filter and display filter, which can be used over live traffic and/or with saved capture files. Filters provide advanced capabilities in performing packet analysis, where a user is able to separate the unwanted stream of packets from the stream of packets for analysis.

Capture filters

Capture filters enable you to capture only traffic that you want to be captured, eliminating an unwanted stream of packets. Capturing packets is a processor-intensive task, and packet analyzers use a good amount of primary memory while they are running.

Packets are only sent to the capture engine if they meet a certain criterion (capture filter expressions). Capture filters do not facilitate advanced filtering options, as in display filters.

The following is a screenshot of the **Capture Options** window dialog:

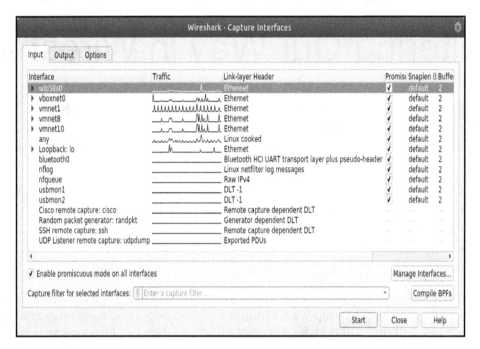

The Capture Options dialog

Let's take a walk through the options available in the **Capture** dialog window:

- **Capture (under input tab)**: Its purpose is to choose which interface you wish to listen on; multiple interfaces can also be selected to run in parallel. The details for every interface are listed under separate columns, such as **Capture**, **Interface**, the name of the interface, whether the promiscuous mode is enabled or not, and so on. Under the **Capture** dialog, you will see a checkbox to toggle the promiscuous mode, which enables you to listen to traffic that is not generated from or headed to your machine.

- **Manage Interfaces**: Facilitates addition or removal of a new interface for listening purposes. You can add even remote machine interfaces to listen remotely.

- **Capture Filter**: Lists capture filters and also facilitates the addition of new user-defined filters:

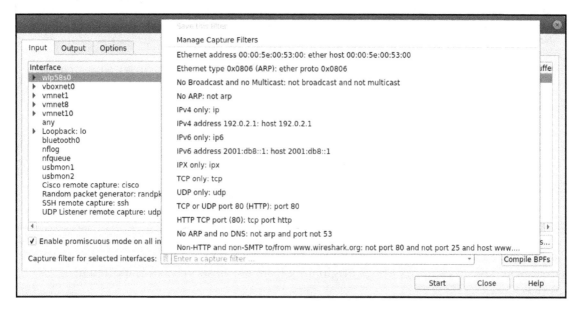

Default Capture Filters

The **Berkley Packet Filtering (BPF)** syntax is an industry standard used for designing filters expressions and is supported by protocol analyzers such as `tcpdump`, which makes a filter's configuration file portable.

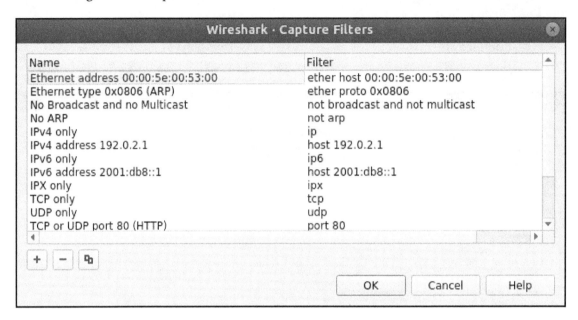

The following are the steps to create your first capture filters expression; consider a scenario where you have to capture packets originating from a web server that is located at `10.10.10.157`:

1. Open the **Capture Options** dialog.
2. Click on **Capture Filter**.
3. Click on **New**.
4. Write `Filtering Host` inside the **Filter name** textbox.

5. Write `host 10.10.10.157` inside the Filter String textbox:

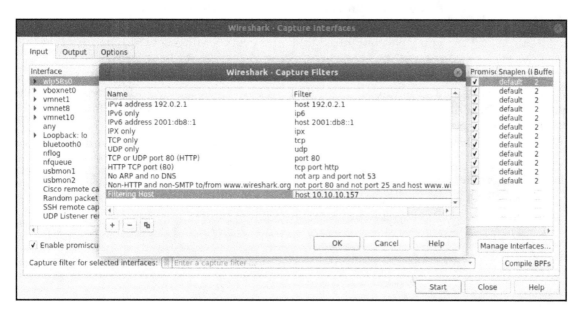

Creating a sample capture filter

6. Once done, click on **OK**; if you've entered everything correctly (mostly the filter expression), the textbox followed by the **Capture Filter** button will be displayed with a green background.

7. **Capture Files (under output tab)**: Use this option to append stream of packets to an existing trace file. The captured packets will be added to the file of your choice. If you haven't chosen any, a temporary file will be created. For more advanced way of saving packets to single/multiple files, try the following:
   - **Create a new file automatically after**: After capturing a certain amount of data (KB, MB or GB), Wireshark will create a new file to save a stream of packets. For instance, I want to create a new file after Wireshark captures 2 MBs of data.
   - **Next File Every (time)**: After a certain amount of time (seconds, minutes, or hours), Wireshark will create a new file to save a stream of packets. For instance, I want to create a new file every five minutes.

- **Ring buffer**: Use this option to set a limit for creation of new files based on the previously mentioned criteria. For example, you have selected the **Ring buffer** option and set the number of files to 5, and you have configured that after every 5 MB, a new file should be created.

According to this configuration, after every 5 MBs of data, a new file will be created and the packets will be written to it. Once the limit that you specified in the **Ring Buffer** is met, Wireshark will not create a new file; instead, it will start saving to the first file and append all captured packets to it. The following screenshot shows a similar kind of configuration:

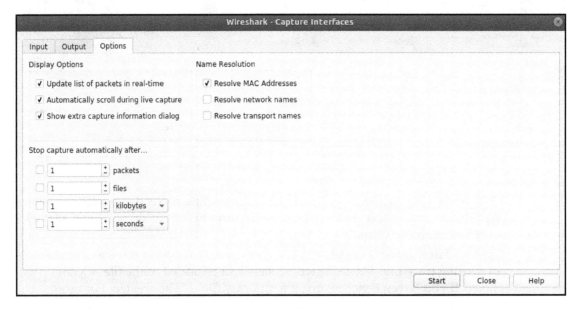

The Capture Files option

- **Stop Capture Settings (options tab)**: This option lets you stop the capturing process after a certain condition is triggered; we have four different kinds of triggers. They are stated as follows:
  - **Packet(s)**: Stop capturing after a certain count of packets is reached
  - **File(s)**: Stop capturing after the creation of a certain number of files
  - **Kilobytes(s)**: Stop capturing after capturing a certain amount of data
  - **Seconds(s)**: Stop capturing after running for a certain period

What if we select more than one option at a time, as shown in the following screenshot:

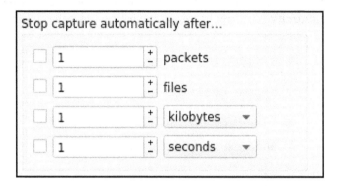

The Stop Capture options

You can activate more than one option at a time; Wireshark will stop capturing whichever condition is met first.

- **Name Resolution (options tab)**: If selected, this feature can resolve the Layer 2, 3, and 4 addresses to their corresponding names:

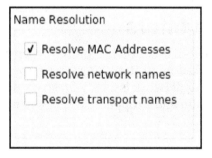

Name Resolution

- **Display Options (options tab)**: Use this option to customize how stream of packets and related information will be show in the **Packet List Pane** and the **Protocol hierarchy** window. Refer to the following screenshot:

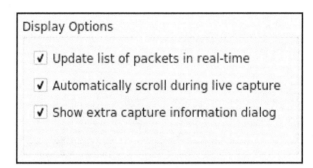

Display Options

- **Update list of packets in real-time**: **Packet List Pane** is updated instantly as soon as a new packet is captured, and the pane will scroll automatically to display the most recent packets

Why use capture filters

Capturing only traffic that meets a criterion is required when a large volume of packets is flowing in network. Creating custom capture filters can come in handy for analyzing a root cause our while troubleshooting network issues. Wireshark discard packets that do not meet the capture filter expression and dropped packets will not be passed to the capturing engine.

How to use capture filters

Use the **Berkley Packet Filter** (BPF) syntax to create capture filters through capture filter dialog.

BPF is a combination of two arguments: identifiers and qualifiers, which are explained as follows:

- **Identifiers**: Search criteria is your identifier. For example, capture filter like `host 192.168.1.1`, where the value `192.168.1.1` is an identifier.
- **Qualifiers**: These are categorized into further three sections:
    - **Type**: There are three types of type qualifiers: `host`, `port`, and `net`. A type qualifier refers to the name or the number that your identifier refers to, e.g. in your capture filter `host 192.168.1.1`, `host` is the type qualifier.
    - **Direction**: Sometimes, when you need to capture packets from a source or destination, specify direction qualifiers along. For example, in the `src host 192.168.1.1` capture filter, `src` specifies to capture packets originating from `192.168.1.1`. Likewise, if you specify `dst host 192.168.1.1`, would capture packets only destined to `host192.168.1.1`.
    - **Proto**: This qualifier is for filtering packets pertaining to a specific protocol. For example, if you want to capture `http` traffic coming from host `192.168.1.1`, then expression will be `src host 192.168.1.1 and tcp port 80`
- Wireshark support usage of *and or* operators to concatenate more than one expressions refer to following examples:
    - Filter `src host 192.168.1.1 and tcp port 80` states that all the packets originating from `192.168.1.1` and going to port `80` should only be captured.
    - Filter `src host 192.168.1.1 or tcp port 80`, states that every packet originating from `192.168.1.1` or any packet associated with port `80` should only be captured.
    - Filter `not port 80` states that any packet associated with port `80` should not be captured.

An example capture filter

To access the default filters, go to **Capture | Capture Filers** or click on the **Capture Options** button in the main toolbar and click on **Capture Filter**.

Refer to the following table for sample capture filters:

Filters	Description
host 192.168.1.1	All traffic associated with host 192.168.1.1
port 8080	All traffic associated with port 8080
src host 192.168.1.1	All traffic originating from host 192.168.1.1
dst host 192.168.1.1	All traffic destined to host 192.168.1.1
src port 53	All traffic originating from port 53
dst port 21	All traffic destined to port 21
src 192.168.1.1 and tcp port 21	All traffic originating from 192.168.1.1 and associated with port 21
dst 192.168.1.1 or dst 192.168.1.2	All traffic destined to 192.168.1.1 or destined to host 192.168.1.2
not port 80	All traffic not associated with port 80
not src host 192.168.1.1	All traffic not originating from host 192.168.1.1
not port 21 and not port 22	All traffic not associated with port 21 or port 22
tcp	All tcp traffic
Ipv6 tcp or udp host www.google.com ether host 07:34:AA:B6:78:89	All ipv6 traffic All TCP or UDP traffic All traffic to and from Google's IP address All traffic associated with the specified MAC address

Display filters

Display filters are flexible and powerful when compared to capture filters. Display filters do not discard any packets; instead, the packets are hidden. Discarding packets is not a very effective practice because, once the packets are dropped, they cannot be recovered. Applying a display filter will limit the packets to be displayed in the list pane of Wireshark.

A display filter can be used for a capture file and live traffic in the **Filter** dialog box located above the **Packet List Pane**. Display filters support variety of arguments such as IP, port, protocol, and so on.

Let's learn how to use the display filter expression dialog for creating filters.

The filter expression

1. Click on the **Expression** button to configure a display filter

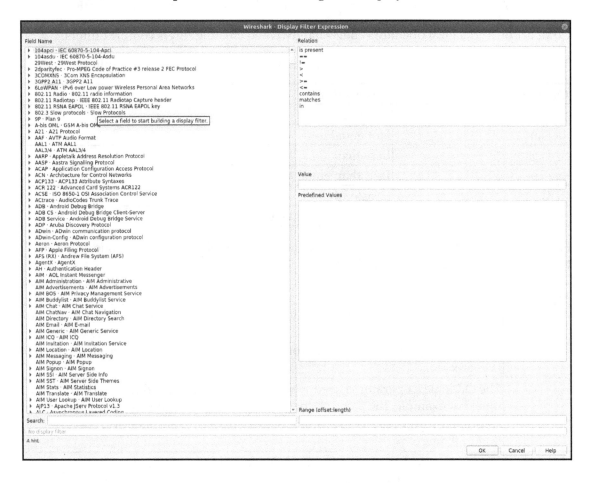

2. For example, if you want to see only packets associated with `ip:192.168.1.1`, then scroll down in the **Field Name** to find **IPv4**. Then, expand the section and choose the **ip.addr** option.

3. From the **Relation** box next to it, choose the operator you wish to add in your expression.

4. At last, write the IP you or in the **Value (IPv4 address)** box and click **OK**

Comparison and logical operators comes handy when creating filters complex filters.

The following table lists the comparison operators that can be used to create filters:

Operator	Description
`==/eq`	Equal to
`!=/ne`	Not equal to
`</lt`	Less than
`<=/le`	Less than equal to
`>/gt`	Greater than
`>=/ge`	Greater than equal to

Following is the list of logical operators that are used to combine more than one criterion together. The following table lists all of them:

Operator	Description
`AND/&&`	The AND logical operator is used when we want both parts of the expression to state `true`. For example, the `ip.src==192.168.1.1` and `tcp` filters would only display packets originated from `ip 192.168.1.1` and associated with the `tcp` protocol.
`OR/||`	The OR logical operator is used when we focus on one condition to be true at a time; For example, the `port 53` or `port 80` filters would display all packets associated with port 53 (`DNS`) along with all packets associated with port `80` (`http`) if any.
`NOT/!`	The NOT logical operator is used when we want to exclude some packets from the list pane. For example, the `!dns` filter would hide all the packets associated with the DNS protocol.

Retaining filters for later use

Retaining filters saves time and effort required to type complex display filters. Wireshark facilitates retaining through saving custom filters. To create one for yourself, following are the steps:

1. Go to **Analyze** | **Display filters**:

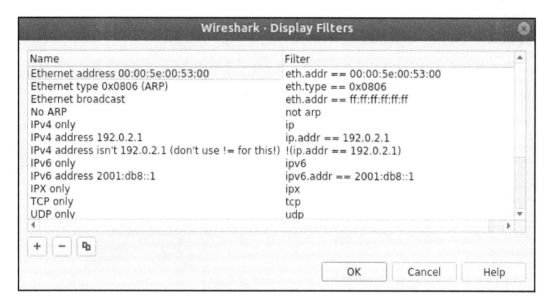

Adding Display Filters

2. Click on **New (+)**, enter the values in the **Filter name** and **Filter string** fields. For instance, we want to create a display filter for NO ARP packets. Then, the values will look like the following screenshot:

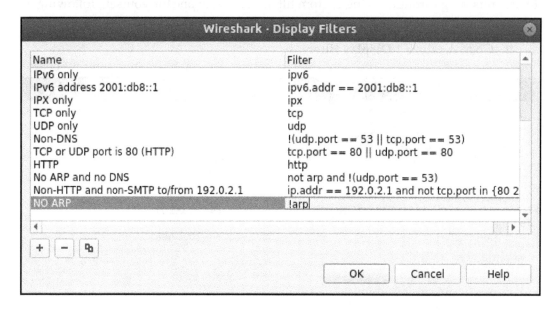

Name	Filter
IPv6 only	ipv6
IPv6 address 2001:db8::1	ipv6.addr == 2001:db8::1
IPX only	ipx
TCP only	tcp
UDP only	udp
Non-DNS	!(udp.port == 53 || tcp.port == 53)
TCP or UDP port is 80 (HTTP)	tcp.port == 80 || udp.port == 80
HTTP	http
No ARP and no DNS	not arp and !(udp.port == 53)
Non-HTTP and non-SMTP to/from 192.0.2.1	ip.addr == 192.0.2.1 and not tcp.port in {80 2
NO ARP	!arp

Creating a new filter

3. Click on **Apply**. Now, your recently created filter will be listed at the bottom of list, which can be used later.

4. Make sure that the **Filter String** box is shown with a green background, which means that your expression is correct; if it is in red color, then something is wrong, and if it is in yellow, this denotes that the results can be unexpected.

5. Click on the **Expression** button next to the **Filter string** box, to create filters through click and selecting what you require.

6. The **Delete (-)** button will delete an existing filter from the list.

7. The **Cancel** button will discard any unsaved changes and close the window.

8. The **Ok** button commits **Save** and closes the window.

Searching for packets using the Find dialog

For searching packets that meets a criterion use the Find tool bar adjacent to display filter. You can access the **Find utility** by navigating to **Edit** | **Find packets** or using the shortcut *Ctrl + F*:

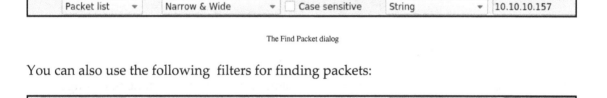

The Find Packet dialog

You can also use the following filters for finding packets:

Let's see some more configurable options available:

- The **display filter**: Find packets based on specific IP /Port/ Protocol, for example:
    - `ip.addr == 192.168.1.1` (based on an IP address)
    - `port 8080` (based on a port number)
    - `http` (based on a protocol)

- The **Hex value**: If you have the hex value for a packet, then use this option. For example, write in the physical address separated by colons, for example:
  - `0A:C4:22:90:45:00`
  - `AA:BB:CC`
- **String**: Enter the name of the DNS server, name of the machine, and any name that you are looking for (enter any string or word), for example:
  - Cisco
  - An administrator
  - A web server
  - Google
- **Search In**: Through this you can search in specific pane of Wireshark. For instance, if you are looking for a packet which matches the value **Google** (the ASCII value in the packet bytes pane will be matched). So, first choose the **String** option and then choose the **Packet bytes** from the first drop down.
- **String Options**: To enable and use this option, first select the **String** option and then select **Case-Sensitive** and then if you want, choose the character width as well.

To move back and forth between the matched packets, you can use *Ctrl + N* (next) and *Ctrl + B* (previous).

Colorize traffic

For better and convenient viewing experience colorization of traffic is done to distinguish between different stream of packets. Colorization helps in differentiating between similar looking packets in ease.

To access the default colorizing profiles navigate to **View** | **Coloring rules** as shown in the following screenshot:

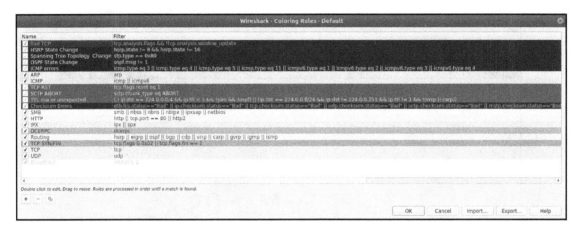

Coloring rules

All rules that are currently saved as part of your global configuration file to colorize traffic are listed in this dialog. Every packet listed in the packet list pane follows the rules defined in Coloring rules windows, which gives them a distinctive look.

Let's use this feature and color the http error packets with a color combination of our choice. Say, for instance, a web server is configured and up and running file sharing purpose. Now, a client is trying to do directory listing and gets HTTP 404 error messages. These error messages are shown in the packet list pane and colored using the default http coloring rule that makes these errors less visible to us. To identify such packets quickly, colorize the HTTP 404 error messages with a black background and with a cyan foreground. Follow the steps to configure the same.

1. Linux box is the client configured on IP `172.16.136.129`, and Macintosh running on `172.16.136.1` that is configured as a web server:

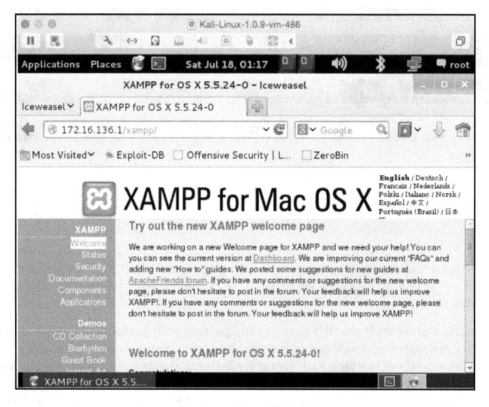

The web server running on 172.16.136.1

2. Normal traffic from a Linux-accessing web server looks something shown as follows:

No.	Time	Source	Destination	Protocol	Length	Info
1	0.000000000	172.16.136.129	172.16.136.1	TCP	60	55658→80 [SYN] Seq=0 Win=2920
2	-950618696.077286000	172.16.136.1	172.16.136.129	TCP	64	80→55658 [SYN, ACK] Seq=0 Ack
3	-2021440336.836621000	172.16.136.129	172.16.136.1	TCP	52	55658→80 [ACK] Seq=1 Ack=1 Wi
4	-1898165200.561362000	172.16.136.1	172.16.136.129	TCP	52	[TCP Window Update] 80→55658
5	41863044.612094000	172.16.136.129	172.16.136.1	HTTP	355	GET /xampp/ HTTP/1.1
6	0.001038000	172.16.136.1	172.16.136.129	TCP	52	80→55658 [ACK] Seq=1 Ack=304
7	0.084997000	172.16.136.1	172.16.136.129	HTTP	940	HTTP/1.1 200 OK (text/html)
8	0.085422000	172.16.136.129	172.16.136.1	TCP	52	55658→80 [ACK] Seq=304 Ack=88
9	381882809.099438000	172.16.136.129	172.16.136.1	HTTP	400	GET /xampp/head.php HTTP/1.1
10	0.106560000	172.16.136.1	172.16.136.129	TCP	52	80→55658 [ACK] Seq=889 Ack=65
11	-1437096632.910449000	172.16.136.129	172.16.136.1	TCP	60	55659→80 [SYN] Seq=0 Win=2920
12	-950618696.095408000	172.16.136.1	172.16.136.129	TCP	64	80→55659 [SYN, ACK] Seq=0 Ack
13	-136085583.409139000	172.16.136.129	172.16.136.1	TCP	52	55659→80 [ACK] Seq=1 Ack=1 Wi
14	-1321431987.061550000	172.16.136.1	172.16.136.129	TCP	52	[TCP Window Update] 80→55659

3. Now that everything is up and running, we will try to do some directory listing manually from client machine, to generate `HTTP 404` error messages.

4. The traffic generated through this request is captured and can be seen in the following screenshot:

No.	Time	Source	Destination	Protocol	Length	Info
92	675.958501000	172.16.136.129	172.16.136.1	TCP	52	55667–80 [ACK] Seq=1 Ack=1
93	-1278177470.593326000	172.16.136.1	172.16.136.129	TCP	52	[TCP Window Update] 80–556
94	675.958885000	172.16.136.129	172.16.136.1	HTTP	362	GET /xampp/abc.jpg HTTP/1.
95	238256651.845389000	172.16.136.1	172.16.136.129	TCP	52	80–55667 [ACK] Seq=1 Ack=3
96	-456584943.391379000	172.16.136.1	172.16.136.129	TCP	657	[TCP segment of a reassemb
97	675.981774000	172.16.136.1	172.16.136.129	TCP	483	[TCP segment of a reassemb
98	675.981788000	172.16.136.1	172.16.136.129	TCP	282	[TCP segment of a reassemb
99	-511200557.945281000	172.16.136.1	172.16.136.129	TCP	273	[TCP segment of a reassemb
100	-1437100881.841330000	172.16.136.1	172.16.136.129	HTTP/XML	60	HTTP/1.1 404 Not Found
101	-1177513788.717358000	172.16.136.129	172.16.136.1	TCP	52	55667–80 [ACK] Seq=311 Ack
102	-1177513788.717358000	172.16.136.129	172.16.136.1	TCP	52	55667–80 [ACK] Seq=311 Ack
103	675.982078000	172.16.136.129	172.16.136.1	TCP	52	55667–80 [ACK] Seq=311 Ack
104	-1177513788.717358000	172.16.136.129	172.16.136.1	TCP	52	55667–80 [ACK] Seq=311 Ack

HTTP 404 Traffic

We can see, in the preceding captured traffic, that the client requested the
`abc.jpg` resource, which was not available; thus, the client received a **404 Not found** error.

5. We figured out easily because there is just one client requesting a single resource. Consider a production environment with thousands of clients. In such cases, coloring a specific set of packets with a different rule is a game changer.

6. Navigate to **Edit Coloring Rules** | **New (+)**. Type **HTTP 404** in the **Name** box.

7. Type `http.response.code==404` in the **Filter** box. Choose the **Foreground Color** option as `Cyan`, and choose the **Background Color** option as **Black**. Then, click on **OK**:

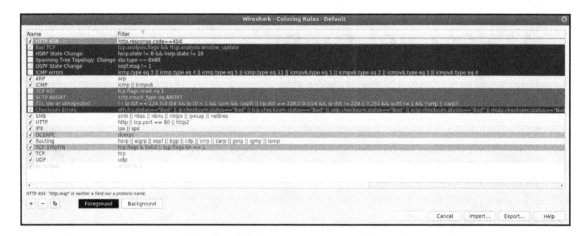

8. Click **OK** and you will see the new rule in action:

No.	Time	Source	Destination	Protocol	Length	Info
94	675.958885000	172.16.136.129	172.16.136.1	HTTP	362	GET /xampp/abc.jpg HTTP/1.1
95	238258651.845389000	172.16.136.1	172.16.136.129	TCP	52	80–55667 [ACK] Seq=1 Ack=31]
96	-456584943.391379000	172.16.136.1	172.16.136.129	TCP	657	[TCP segment of a reassemble
97	675.981774000	172.16.136.1	172.16.136.129	TCP	483	[TCP segment of a reassemble
98	675.981788000	172.16.136.1	172.16.136.129	TCP	282	[TCP segment of a reassemble
99	-511200557.945281000	172.16.136.1	172.16.136.129	TCP	273	[TCP segment of a reassemble
100	-1437100881.841330000	172.16.136.1	172.16.136.129	HTTP/XML	60	HTTP/1.1 404 Not Found
101	-1177513788.717358000	172.16.136.129	172.16.136.1	TCP	52	55667–80 [ACK] Seq=311 Ack=6
102	-1177513788.717358000	172.16.136.129	172.16.136.1	TCP	52	55667–80 [ACK] Seq=311 Ack=]
103	675.982078000	172.16.136.129	172.16.136.1	TCP	52	55667–80 [ACK] Seq=311 Ack=]
104	-1177513788.717358000	172.16.136.129	172.16.136.1	TCP	52	55667–80 [ACK] Seq=311 Ack=]
105	-1437162184.138035000	172.16.136.129	172.16.136.1	TCP	52	55667–80 [ACK] Seq=311 Ack=]

After applying the new coloring rule

Coloring rules are applied to the packet list pane in a top-to-bottom manner. With every packet, there is coloring rule information attached that can be listed from **Packet Details Pane** under the **Frame** section, as shown as follows:

Coloring info in a frame header

Create new Wireshark profiles

Profiles are like customized virtual environments, which saves significant amount of time while auditing/troubleshooting a network.

To create a profile, follow these steps:

1. Right-click on the **Profile** column in **Status Bar (bottom right corner of window):**

Profile: Default

2. Click on + in the pop-up dialog:

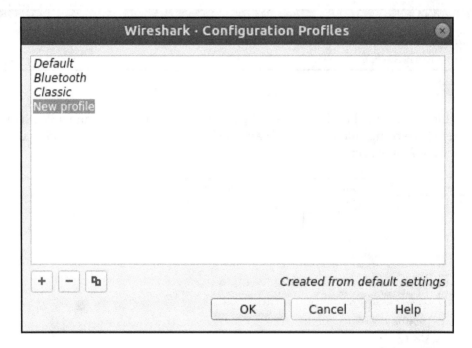

3. Now, choose any profile you wish to use as a template (if any) and type the name of the new profile.

4. And then, click on **OK**.

Now, in the status bar, you will see the new profile has been activated. The changes that you will make in this profile stays here, for example, you create capture/display filters, change protocol preferences, and change color preferences, and so on.

```
Profile: New profile
```

Also, importing and exporting profiles is easy just copy and paste the Profile configuration files in a Wireshark directory to use.

Summary

Filtering traffic lets you capture and see only stream of packets you want; there are two types of filters: display filters and capture filters.

Display filters hide the packets; however, capture filters discard the packets that do not meet user defined expression and discarded packets are not passed to the capturing engine.

Capture filters use the BPF syntax, which is an industry standard and is used by several other protocol analyzers.

Find utility is useful and can be accessed from the Edit menu in Wireshark. The Find utility gives various vectors to search a packet(s) and related details.

Coloring preferences comes handy when filtering a set of traffic. Distinguishing packets becomes easy, as the matched packets will be displayed with a unique coloring scheme.

Profiles are like virtual scenarios that saves time and efforts. Changes made to a profile with respect to display/capture filter and color/protocol/time preferences, stays within the same.

4
Analyzing Application Layer Protocols

This chapter will help you understand the approach and methodology for analyzing application layer protocols such as HTTP, SMTP, FTP, and DNS through Wireshark. Application layer protocols typically interfaces between a client and server.

It is critical to understand the structure of application layer protocol packets in order to identify anomalies efficienctly. We will be discussing the following topics in detail throughout this lesson:

- Analysis of common application layer protocols
- Assembling VoIP packets
- Decrypting encrypted traffic

Domain Name System (DNS)

Imagine a world of internet where you have to type a random numerical value (IP address) in your web browser's address bar, instead of a name, to visit a website. Also, imagine that each numerical figure is different. Considering this, how many numbers (IP addresses) can you memorize? 5? 10? Perhap, 50 at max? So, now, you are confined to visiting just 50 websites.

For the sake of a limitless web experience, DNS comes to our rescue. DNS stores a dataset (zone file) of website names mapped to their current IP addresses, along with the names of the domains. Each entry in the zone file is termed a resource record (combination of website name and its IP). DNS uses TCP and UDP, both for different purposes, over the port 53 by default.

How does DNS work? So, as a client, when you try to visit a website from a browser, your request (DNS query) is sent to an internal DNS server (if any) that looks up the resource records it contains. If the DNS server knows the IP address for the domain you are trying to visit, your PC will get a reply (DNS response) containing the IP address of the website you desire to visit, else your query will be forwarded to external DNS servers on the web (for example, google DNS servers at 8.8.8.8, 4.4.2.2, and so on.).

Dissecting a DNS packet

A DNS packet consists of multiple fields that are briefly discussed here:

- **Transaction ID**: This is a number that keeps track of a domain query and it's corresponding response.
- **Query/response**: Every DNS packet is marked as a query or a response.
- **Flag bits**: Each query and response contains a different set of flag bits, which are as follows:
    - **Response**: The message is a query or a response.
    - **Opcode**: This determines the type of query contained. The Opcode ranges between 0-15. Refer to the following table:

0	1	2	3	4	5	6-15
Standard query	Inverse query	Server status request	Unassigned	Notify	Update	Unassigned

- **Truncated**: This determines whether the packet is truncated if its size is large (greater than 512 bytes).
- **Recursion desired**: The query sent by your client is supposed to go on a recursive search procedure from one DNS server to another if the resource record you are looking for is not present in the primary DNS.
- **Recursion available**: If this bit is set, then it means the recursion that your client requested is available.
- **Reserved (z)**: As defined by RFC 1035; reserved for future use, must be set to zero for all queries and responses.

- **Response code:** The values in this field signifies the response. This field is used to signify whether there are errors and the types of errors. Here are the possible code values that you can receive:

0	1	2	3	4	5
No error	Format error	Server failure	Name error	Not implemented	Refused

- **Questions**: Number of queries present in the packet.
- **Answers**: Number of answers sent in response to the query.
- **Authority RRs**: Number of authority resource records sent as response.
- **Additional RRs**: Number of additional resource records sent as response.
- **Query section**: The query sent to the DNS server; it should be the same in the response received.
- **Answer section**: Answer consists of the resource records that came in as response.
- **Type**: Type of query sent. Refer to the following table for common query types:

A	NA	MX	SOA	PTR	AAAA	AXFR	IXFR
Host address	Name server	Mail exchange	Start of zone authority	Pointer record	IPv6 address	Full zone transfer	Incremental zone transfer

- **Additional info**: This field includes additional info containing resource records. It is not required to answer the query.

Dissecting DNS query/response

Let's consider a scenario to understand the way DNS works. A client sends a query to a DNS server that possesses name resolution information. Using this information, the client can start IP-based communication. Sometimes, the information the client is looking for is not available with the DNS server it requested. In such cases, the DNS server itself transfers the query to any neighbor DNS it knows about, if recursion is desirable. Refer to the following screenshot, where a request is sent to visit `https://www.google.co.in`. A request from a client located at `192.168.1.103` is sent to the default gateway at `192.168.1.1`. This gateway will forward the query to a DNS server it knows about:

```
▷ Frame 9: 74 bytes on wire (592 bits), 74 bytes captured (592 bits) on interface 0
▷ Ethernet II, Src: Apple_b9:53:ec (d8:bb:2c:b9:53:ec), Dst: Zte_07:73:6c (d0:5b:a8:07:73:6c)
▷ Internet Protocol Version 4, Src: 192.168.1.103 (192.168.1.103), Dst: 192.168.1.1 (192.168.1.1)
▷ User Datagram Protocol, Src Port: 65382 (65382), Dst Port: 53 (53)
▽ Domain Name System (query)
    [Response In: 10]
    Transaction ID: 0x2b4a
  ▷ Flags: 0x0100 Standard query
    Questions: 1
    Answer RRs: 0
    Authority RRs: 0
    Additional RRs: 0
  ▽ Queries
    ▽ www.google.com: type A, class IN
        Name: www.google.com
        [Name Length: 14]
        [Label Count: 3]
        Type: A (Host Address) (1)
        Class: IN (0x0001)
```

DNS query

You may notice that DNS is using UDP as an underlying protocol. If you want to know more about the DNS query being generated, just expand the **Flags** section. This section will list various details, such as whether recursion is available, whether recursion is desired, and what the response code is. Please refer to the following screenshot:

```
▽ Flags: 0x0100 Standard query
    0... .... .... .... = Response: Message is a query
    .000 0... .... .... = Opcode: Standard query (0)
    .... ..0. .... .... = Truncated: Message is not truncated
    .... ...1 .... .... = Recursion desired: Do query recursively
    .... .... .0.. .... = Z: reserved (0)
    .... .... ...0 .... = Non-authenticated data: Unacceptable
```

Expanded flags section

The expanded **Flags** section tells us that the type of DNS packet is a query, the packet data is not truncated, and recursion is desirable if available.

In response to this query, you will observe one packet with the same transaction ID that denotes the association of a DNS query sent by the client. The response for the query will usually consist of an IP address for the domain visited. The requesting machine will be returned a single IP, or maybe multiple IPs available to it. If the domain we are looking for is not available, then it's probable CNAMEs will be returned in as favor.

Refer to the following screenshot to understand this:

```
▷ Frame 10: 154 bytes on wire (1232 bits), 154 bytes captured (1232 bits) on interface 0
▷ Ethernet II, Src: Zte_07:73:6c (d0:5b:a8:07:73:6c), Dst: Apple_b9:53:ec (d8:bb:2c:b9:53:ec)
▷ Internet Protocol Version 4, Src: 192.168.1.1 (192.168.1.1), Dst: 192.168.1.103 (192.168.1.103)
▷ User Datagram Protocol, Src Port: 53 (53), Dst Port: 65382 (65382)
▽ Domain Name System (response)
    [Request In: 9]
    [Time: 0.004678000 seconds]
    Transaction ID: 0x2b4a
  ▷ Flags: 0x8180 Standard query response, No error
    Questions: 1
    Answer RRs: 5
    Authority RRs: 0
    Additional RRs: 0
  ▷ Queries
  ▽ Answers
    ▷ www.google.com: type A, class IN, addr 173.194.36.84
    ▷ www.google.com: type A, class IN, addr 173.194.36.83
    ▷ www.google.com: type A, class IN, addr 173.194.36.82
    ▷ www.google.com: type A, class IN, addr 173.194.36.80
    ▷ www.google.com: type A, class IN, addr 173.194.36.81
```

DNS response

As I said, we could get multiple replies. If you notice the **Answer RRs** section, we have received five replies for the www.google.com domain. For verification that the response received belongs to the previous query only, just match the **Transaction ID**.

Expand any section in the **Answers** category to view more details. Refer to the following screenshot:

```
▽ Answers
    ▽ www.google.com: type A, class IN, addr 173.194.36.84
        Name: www.google.com
        Type: A (Host Address) (1)
        Class: IN (0x0001)
        Time to live: 13
        Data length: 4
        Address: 173.194.36.84 (173.194.36.84)
```

File transfer protocol

Since the internet came into existence, we have been working with the **file transfer protocol** (**FTP**). FTP uses TCP over port 21 or 20 (by default) to initiate and transfer files over a designated channel. There are only two types channel command channel (port 21) and data channel (port 20). The command channel is used to send and receive the commands and their responses. The data channel is used to send and receive data between the client and the server. However, you will observe random port numbers used to transfer TCP data segments from your client machine.

Dissecting FTP communication packets

There are two types of mode a client can use to communicate with a server: active and passive. In earlier versions of FTP server applications, active mode was enabled by default, but in the latest versions of FTP server applications, passive mode is enabled by default. For understanding these modes in detail, let's use the following scenario.

Let's say an FTP server is configured at IP 172.16.136.129 and a client at IP 172.16.136.1.

Typically, every request sent from the client is a specific command set, to which the server responds with a numerical value followed by a text message. See the following screenshot for reference followed by a short analysis:

```
 4 0.018723000            172.16.136.129   172.16.136.1     FTP   88 Response: 220 Welcome to Charit's FTP se
 5 555032032.287455000    172.16.136.1     172.16.136.129   TCP   52 56982→21 [ACK] Seq=1 Ack=37 Win=131728 L
 6 -952210303.718297000   172.16.136.1     172.16.136.129   FTP   62 Request: USER abc
 7 -143593220.746255000   172.16.136.1     172.16.136.1     TCP   52 21→56982 [ACK] Seq=37 Ack=11 Win=29696 L
 8 4.629189000            172.16.136.129   172.16.136.1     FTP   86 Response: 331 Please specify the passwor
 9 4.629206000            172.16.136.1     172.16.136.129   TCP   52 56982→21 [ACK] Seq=11 Ack=71 Win=131696
10 5.732635000            172.16.136.1     172.16.136.129   FTP   62 Request: PASS abc
11 -1086390884.249094000  172.16.136.129   172.16.136.1     FTP   75 Response: 230 Login successful.
12 2070317539.792672000   172.16.136.1     172.16.136.129   TCP   52 56982→21 [ACK] Seq=21 Ack=94 Win=131672
```

The server requested the password, which the client provided. Once the server receives and validates the password, the user will be logged in. In our case, the password is correct, so the client receives 230 as a response code followed by a Login Successful message.

Commands issued from the client side can have arguments or no arguments, and the data transmitted between the devices can be seen in the TCP header of the packet, as shown here:

```
43 -544276953.032968000   172.16.136.1     172.16.136.129   FTP    58 Request: LIST
44 894485615.992341000    172.16.136.129   172.16.136.1     TCP    60 20→57197 [SYN] Seq=
45 894485615.992407000    172.16.136.1     172.16.136.129   TCP    64 57197→20 [SYN, ACK]
46 894485615.992662000    172.16.136.129   172.16.136.1     TCP    52 20→57197 [ACK] Seq=
47 894485615.992690000    172.16.136.1     172.16.136.129   TCP    52 [TCP Window Update]
48 -540049189.689031000   172.16.136.129   172.16.136.1     FTP    91 Response: 150 Here
49 894485615.993039000    172.16.136.1     172.16.136.129   TCP    52 57196→21 [ACK] Seq=
50 894485615.993493000    172.16.136.129   172.16.136.1     FTP-DATA  314 FTP Data: 262 byte
51 3493485A8 2290390000   172 16 136 1     172 16 136 129   TCP    52 57197 20 [ACK] Seq=
```

```
▷ Frame 50: 314 bytes on wire (2512 bits), 314 bytes captured (2512 bits) on interface 0
▷ Raw packet data
▷ Internet Protocol Version 4, Src: 172.16.136.129 (172.16.136.129), Dst: 172.16.136.1 (172.16.136.1)
▷ Transmission Control Protocol, Src Port: 20 (20), Dst Port: 57197 (57197), Seq: 1, Ack: 1, Len: 262
  FTP Data (drwxr-xr-x   2 1001    1002        4096 Aug 03 00:45 Desktop\r\n-rw-r--r--    1 0
```

FTP-data returned

Frame 43 shows that the client issued the LIST command, which was processed by the server, and that 262 bytes of data was returned. FTP-based communication can be seen in plaintext through protocol analyzers, which is also a weakness often exploited.

Reassembling the FTP data stream is easy because apart from the data, there is nothing that is transmitted. There is no code or command that gets appended to the packets. To reassemble the TCP stream of FTP packets, just right-click on the selected packet and choose the **Follow TCP Stream** option to view.

Refer to the following screenshot:

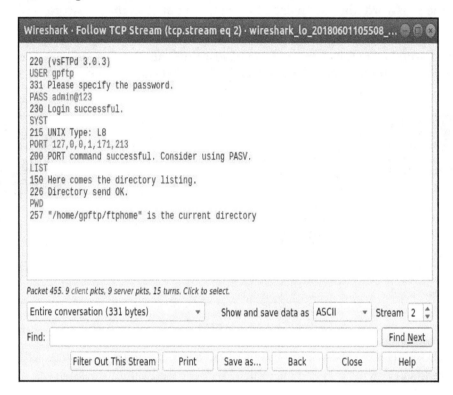

FTP stream

The entire communication between the client and the server that happened over the data and command channels is translated into human-readable format. Text in red is what the client sent, and text in blue is what the client received. It is recommended to use secure versions of FTP in order to mitigate the vulnerability.

Hypertext Transfer Protocol (HTTP)

Data on the web is transferred using the HTTP/HTTPS application layer protocol. Normal communication in HTTP follows a request/response model, where the communication between a client and a server is coordinated by a set of rules. The client requests for a certain resource to the server and then receives a status code that specifies the current status of the requested resource. If available then, the resource is also sent along with the status code, else the client would receive a not-available status code.

How request/response works

Web servers utilize HTTP to serve web pages to the requesting clients. At the beginning of every HTTP session, the TCP three-way handshake takes place. It creates a dedicated channel between the communicating hosts followed by HTTP and data packets, which are sent in and received while the session is active. For instance, say you are visiting a web server located at `http://172.16.136.129` from a client at `172.16.136.1`. Using our client-server infrastructure, we will try to capture the requests sent and responses received.

I will try to visit the home page located at the server mentioned earlier and will capture the traffic generated for the whole session; that is, the requests sent and responses received. Take the following steps to replicate the scenario.

Request

Following are the steps for the preceding scenario:

1. Open your browser and type the **Uniform Resource Locator** (**URL**) of any website.
2. The website is located at http://172.16.132.129 (a local web server). Here is the screenshot for your reference:

3. The following screenshot depicts the packets captured as a result of visiting the web server:

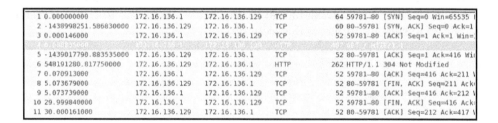

4. All these packets get generated as soon as you press *Enter*. As you can see, the first three packets are TCP three-way handshake packets where our client is requesting that the server creates a dedicated channel. However, if the server daemon wasn't running or the server wasn't accepting our requests, for some reason then we would have seen RST ACK packets, like the one shown here:

5. This error states that the server is out of service or is not supposed to respond to our requests (firewalled or restricted zone).

6. After the TCP packets, the first HTTP request sent by our client is observed. Every request comprises a couple of elements that are sent to the server:

```
GET / HTTP/1.1\r\n
Host: 172.16.136.129\r\n
If-None-Match: "12625d-bc-51c6ab45063d1"\r\n
Accept: text/html,application/xhtml+xml,application/xml;q=0.9,*/*;q=0.8\r\n
If-Modified-Since: Mon, 03 Aug 2015 16:31:40 GMT\r\n
User-Agent: Mozilla/5.0 (Macintosh; Intel Mac OS X 10_10_3) AppleWebKit/600.6.3
Accept-Language: en-us\r\n
Accept-Encoding: gzip, deflate\r\n
Connection: keep-alive\r\n
```

HTTP request

- In the first line, there are three things passed on to the server as the arguments, which are the HTTP method, the requested resource, and the location / (root directory).

- The Host argument is required by the HTTP/1.1 protocol requests. The value of this field is the web server's address that you typed in the address bar of the browser.

- The ACCEPT parameter specifies what kind of content is acceptable by the requesting client response.

- The If-modified-since parameter is sent from the client to the server, which includes the date and time of your previous request made to the server. If the server contents have been changed since your previous request, then you will receive the new updated page. Otherwise, your system will present you with the locally cached page.

- The user-agent specifies the browser-related information that you are using. This information is to be used by the server to present you with browser-compatible content.

- Parameters such as **Accept-Language** and **Accept-Encoding** are passed on to the server to inform us of what type of content is acceptable to the client.

- The **Connection**-alive parameter specifies whether the client wishes to keep the connection working after this particular request has been processed.

Response

1. After the fourth packet, the server acknowledges the client's request to get to the web server's root directory. The server starts transmitting the resource that the client requested.

2. The sixth packet in the list pane is what the client received, a status code followed by a short message, including the content of the resource requested. Refer to the following screenshot illustrating the HTTP response:

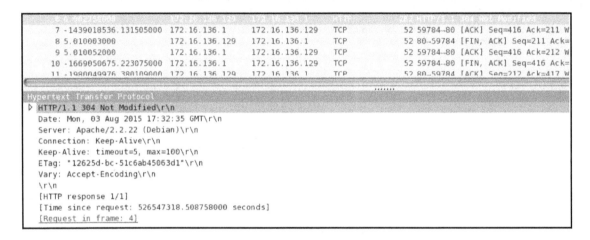

```
7 -1439018536.131505000  172.16.136.1    172.16.136.129  TCP  52 59784-80 [ACK] Seq=416 Ack=211 W
8 5.010003000            172.16.136.129  172.16.136.1    TCP  52 80-59784 [FIN, ACK] Seq=211 Ack=
9 5.010052000            172.16.136.1    172.16.136.129  TCP  52 59784-80 [ACK] Seq=416 Ack=212 W
10 -1669050675.223075000 172.16.136.1    172.16.136.129  TCP  52 59784-80 [FIN, ACK] Seq=416 Ack=
11 -1980049976.380109000 172.16.136.129  172.16.136.1    TCP  52 80-59784 [ACK] Seq=212 Ack=417 W
```

```
Hypertext Transfer Protocol
▷ HTTP/1.1 304 Not Modified\r\n
  Date: Mon, 03 Aug 2015 17:32:35 GMT\r\n
  Server: Apache/2.2.22 (Debian)\r\n
  Connection: Keep-Alive\r\n
  Keep-Alive: timeout=5, max=100\r\n
  ETag: "12625d-bc-51c6ab45063d1"\r\n
  Vary: Accept-Encoding\r\n
  \r\n
  [HTTP response 1/1]
  [Time since request: 526547318.508758000 seconds]
  [Request in frame: 4]
```

HTTP response

3. As a part of TCP communication, the client will acknowledge every packet sent by the server, as seen in the seventh packet.

4. Let's dissect the response elements for packet number six:
   - The first line consists of three arguments sent in response. They denote the HTTP
     protocol version in use, the status code (304 in our case, which specifies
     that the requested resource did not change since the time mentioned in the Date parameter), and finally, a brief description of the status code (not modified in our case).
   - In the third line, the **Server** parameter mentions the name and version of
     the web server. We can see that Apache/2.2.22 is the server that is located at 172.16.136.129.

- The fourth and fifth lines state that the server wishes to keep the connection alive. The duration for which the server wishes to do so is also mentioned in the next line of the parameters.

Simple Mail Transfer Protocol (SMTP)

SMTP is used widely to send and receive emails over a small network. The protocol uses the Sender-SMTP process to send emails and the Receiver-SMTP process to receive emails. This makes SMTP a client-server-based protocol that runs over port 25.

Typically, an SMTP channel for mail transfer is created through a successful TCP three-way handshake followed by a series of SMTP packets:

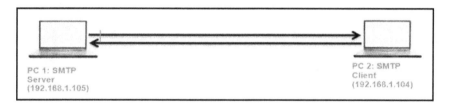

In our lab, we have an SMTP server configured at IP 192.168.1.105 and a client at IP 192.168.1.104. The client will request the server to sends an email to an address known to the client. The server will respond to this request with numerical code, followed by a brief response parameter.

Dissecting SMTP communication packets

Using the Netcat client from a Kali Linux machine, I will connect to the SMTP mail service running on a Windows machine. After a successful three-way handshake, the server will respond with numerical codes with a short summary. Follow these steps to the send an email using command line:

1. Open a connection with the mail server using netcat nc -nv 192.168.1.105 25.
2. Initialize an SMTP session with the HELO testmail command.
3. Specify the from address using the MAIL FROM:<abc@charit.com> command.
4. Specify the recipient's address using the RCPTS TO:<efg@charit.com> command.

5. To enter data into the mail body, type DATA, press *Enter, and type* . *(full-stop; this is a* terminating character, and you can use any character of your choice) Now, type the message you wish to send. Once you are finished typing your mail, type a . (full stop) to mark the ending and press *Enter*.

6. Now, your message will be sent.

The process will generate a couple of packets that contain details about our session. All of these commands mentioned will only work when the server is configured to permit clear text message communication without any authentication; refer to the following screenshot:

1 0.000000000	192.168.1.104	192.168.1.105	TCP	60 57073-25 [SYN] Seq=0 Win=29200 Len=0 MSS
2 1439081651.426767000	192.168.1.105	192.168.1.104	TCP	60 25-57073 [SYN, ACK] Seq=0 Ack=1 Win=1638
3 -41448.227586000	192.168.1.104	192.168.1.105	TCP	52 57073-25 [ACK] Seq=1 Ack=1 Win=29696 Len
4 4205130.997054000	192.168.1.105	192.168.1.104	SMTP	90 S: 220 Charit's.com ESMTP server ready.
5 1439081652.143751000	192.168.1.104	192.168.1.105	TCP	52 57073-25 [ACK] Seq=1 Ack=39 Win=29696 Ler
6 -287363963.384218000	192.168.1.104	192.168.1.105	SMTP	61 C: helo abc
7 1744899513.488830000	192.168.1.105	192.168.1.104	SMTP	82 S: 250 Charit's.com Hello, abc.
8 1439081657.529807000	192.168.1.104	192.168.1.105	TCP	52 57073-25 [ACK] Seq=10 Ack=69 Win=29696 Le
9 1744901809.636462000	192.168.1.104	192.168.1.105	SMTP	79 C: mail from:<abc@charit.com>
10 1744899513.488830000	192.168.1.105	192.168.1.104	SMTP	81 S: 250 Sender OK - send RCPTs.
11 1439081671.468558000	192.168.1.104	192.168.1.105	TCP	52 57073-25 [ACK] Seq=37 Ack=98 Win=29696 Le
12 1439081686.949708000	192.168.1.104	192.168.1.105	SMTP	78 C: rcpts to:<efg@charit.com>
13 4206566.333758000	192.168.1.105	192.168.1.104	SMTP	91 S: 250 Recipient OK - send RCPT or DATA.
14 1439081687.064346000	192.168.1.104	192.168.1.105	TCP	52 57073-25 [ACK] Seq=63 Ack=137 Win=29696
15 1439081688.805525000	192.168.1.104	192.168.1.105	SMTP	57 C: data
16 4207044.779326000	192.168.1.105	192.168.1.104	SMTP	91 S: 354 OK, send data, end with CRLF.CRLF
17 2122359292.356797000	192.168.1.104	192.168.1.105	TCP	52 57073-25 [ACK] Seq=68 Ack=176 Win=29696
18 1439081690.221834000	192.168.1.104	192.168.1.105	SMTP	55 C: DATA fragment, 3 bytes
19 1439081690.447264000	192.168.1.104	192.168.1.105	SMTP	58 C: DATA fragment, 1 byte C: DATA fragment
20 1439081690.454208000	192.168.1.105	192.168.1.104	TCP	52 25-57073 [ACK] Seq=176 Ack=71 Win=16314
21 1439081690.455528000	192.168.1.105	192.168.1.104	TCP	64 [TCP Dup ACK 20#1] 25-57073 [ACK] Seq=17
22 168258645.511998000	192.168.1.104	192.168.1.105	SMTP	54 C: DATA fragment, 2 bytes
23 419451065.438925000	192.168.1.105	192.168.1.104	SMTP	75 S: 250 Data received OK.
24 1439081690.858935000	192.168.1.104	192.168.1.105	TCP	52 57073-25 [ACK] Seq=73 Ack=199 Win=29696
25 168257924.091710000	192.168.1.104	192.168.1.105	SMTP	57 C: DATA fragment, 5 bytes
26 1439081694.129351000	192.168.1.105	192.168.1.104	SMTP	95 S: 221 Charit's.com Service closing chann
27 850006670.085950000	192.168.1.105	192.168.1.104	TCP	52 25-57073 [FIN, ACK] Seq=242 Ack=78 Win=1
28 850006670.085950000	192.168.1.104	192.168.1.105	TCP	52 57073-25 [ACK] Seq=78 Ack=242 Win=29696

SMTP session

Packets from 1-3 are TCP-handshake packets. The handshake is happening between the client and the server. In the fourth packet, the client receives a message stating 220 as the response code. This means the server is available and ready to respond to the client's request. In the sixth packet, the client initializes the standard SMTP session using the HELO command, followed by the sender's and recipient's email addresses, which were confirmed to be correct by the server, with response code 250 in packets 10 and 13. Then there's the email body packet using the DATA command, which was successfully received by the server in packet 23. In the end, the user gracefully closes the connection by issuing the QUIT command, which the server confirmed in packet 26, thus sending FIN, ACK.

Session Initiation Protocol (SIP) and Voice Over Internet Protocol(VOIP)

SIP is a part of the VOIP family, which is a signaling protocol used to create, manage, and terminate VOIP sessions in a networking environment. Examples of SIP include a two-way phone call or a conference call, or multimedia sessions with multiple hosts. After the initiation of the session, the data is transferred through the **Real time Transport Protocol** (**RTP**) over the dedicated channel. Basically, the family of RTPs governs the transport and the flow control of all multimedia items (RTCP controls the flow).

Wireshark can assemble a stream of RTP packets in order to play back the conversation that happened between two parties (use it ethically!).

SIP runs over UDP and commonly uses port 5060. SIP provides us with different call-managing features, such as initiating calls, disconnecting calls, adding someone to a conference call, and transferring calls, though SIP is not going to help you maintain the quality of calls.

Let's discuss the typical VoIP infrastructure through the following diagram. There are three nodes: two of them are clients and one is the IP telephony server, which enables voice communication:

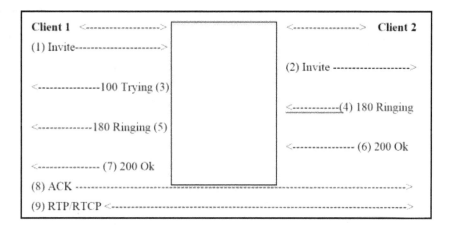

1. **Client 1** sends an **Invite** request to initiate the session using SIP.
2. The telephony server transfers the request to **Client 2**.

2. The telephony server acknowledges **Client 1** with the **100 Trying** packet.

3. **Client 1** receives a **180 Ringing** packet as soon as **Client 2** starts ringing. When **Client 2** on the other side receives the call, it sends the **200 OK** packet, which is forwarded to **Client 1**.

4. Now the client sends the **ACK** packet to acknowledge the receipt of the **200 OK** packet.

5. Now both parties are connected with a dedicated channel, over which the RTP/RTCP packets start flowing back and forth.

6. To end the communication, there will be a **BYE** packet sent by one of the communicating hosts, which is acknowledged by the other end.

7. All of the packets will be sent back and forth between client 1 and 2, due to information only known to telephony server.

8. Once the channel created, all the packets will be sent and received directly by the clients without the server's intervention.

For illustration purposes, I have configured a small VoIP telephony infrastructure using Asterisk PBX that can be downloaded for free. So, our VOIP server is located at `192.168.1.107`, client 1 at `192.168.1.104`, and client 2 at `192.168.1.107`. I am also using an X-lite calling application to call client 2 from client 1. The following is a screenshot of traffic captured in the list pane of Wireshark:

5 0.001673000	192.168.1.107	192.168.1.104	SIP	515 Status: 100 Trying |
172 0.085903000	192.168.1.107	192.168.1.106	SIP/SDP	917 Request: INVITE sip:101@192.168.1.106:5621
177 0.087461000	192.168.1.107	192.168.1.104	SIP	531 Status: 180 Ringing |
178 0.652323000	192.168.1.106	192.168.1.107	SIP	348 Status: 100 Trying |
179 0.959210000	192.168.1.106	192.168.1.107	SIP	501 Status: 180 Ringing |
182 0.961010000	192.168.1.107	192.168.1.104	SIP	531 Status: 180 Ringing |
186 3.827648000	192.168.1.106	192.168.1.107	SIP/SDP	782 Status: 200 OK |
188 3.829335000	192.168.1.107	192.168.1.106	SIP	489 Request: ACK sip:101@192.168.1.106:56215;r
205 3.834786000	192.168.1.107	192.168.1.104	SIP/SDP	820 Status: 200 OK |
211 3.839764000	192.168.1.104	192.168.1.107	SIP	482 Request: ACK sip:101@192.168.1.107 |
1644 10.852745000	192.168.1.104	192.168.1.107	SIP	641 Request: BYE sip:101@192.168.1.107 |
1645 10.853115000	192.168.1.107	192.168.1.104	SIP	489 Status: 200 OK |
1652 10.854002000	192.168.1.107	192.168.1.106	SIP	527 Request: BYE sip:101@192.168.1.106:56215;r
1690 11.042924000	192.168.1.106	192.168.1.107	SIP	467 Status: 200 OK |

SIP traffic

One thing you should consider is placing the analyzer as close as possible to the telephony server so that it will be able to capture every last packet. While capturing, if you cannot see any SIP packets, then you won't be able to capture VOIP packets as well.

Reassembling packets for playback

Yes, it is possible to assemble the VOIP packets back to listen to either side, or both sides, of communication. Let's suppose I want to listen to the message client 1 at IP `192.168.1.104` sent to client 2 at IP `192.168.1.107`. We can use the **Telephony** menu in Wireshark to reassemble the packets and choose the **VOIP Calls** option from the list. The following screenshot illustrates the resulting dialog:

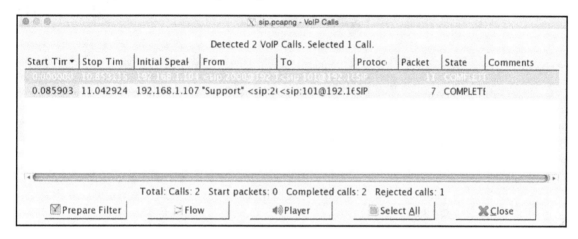

VOIP Calls dialog

Now choose which side of communication you want to listen to. Then click on the **Player** button and configure **Jitter** (Jitter is the variance in packet rate at which the packets are being sent and received. If jitter is high, then there is a chance that your network is dealing with congestion. Calls with high jitter values are not feasible to listen to) and **Time** as illustrated, and click on **Decode**:

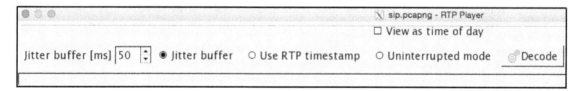

Player dialog

I did not change the default value and clicked directly on the **Decode** button, which reassembled all the VoIP packets for the side of communication I chose, as shown in the following screenshot:

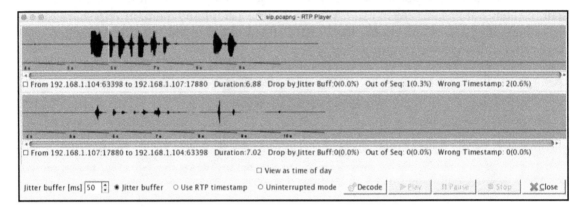

RTP Player

If you want to play the message, check the box just below the scrollbar and click on **Play**. Use this feature for ethical purposes only.

Decrypting encrypted traffic (SSL/TLS)

Yes, it is also possible to decrypt your online TLS traffic into a plaintext SSL stream using Wireshark. Google Chrome and Firefox look for a log file, which stores the TLS session keys. Follow these steps to decrypt a session of encrypted traffic:

1. Create an environment variable with the name SSLKEYLOGFILE that will point to a text file. Your browser will look for this file every time it starts up. To create environment variables, right-click on **My Computer** and go to **Advanced Settings** | **Environment Variables** | **New** | **Specify Name**.
   Enter SSLKEYLOGFILE and **Value:** C:/Users/username/sslkeylog.txt, and click on **OK**.
2. I have created a blank text file, C:/Users/username/sslkeylog.txt (make your new environment variable point to this file).
3. Now open your browser and visit a website enabled with TLS/SSL.
   For demonstration purpose, I have my own SSL web server located at 192.168.1.106 using a client located at 192.168.1.105:

4. After you visit any secure website enabled with SSL, your `sslkeylog.txt` will be populated with some random numbers, as shown in the following screenshot. If not, cross check your settings before moving on:

```
CLIENT_RANDOM 17999a56ea29e69bcb242b441b1b519e
0b3b16e79b9a46bfdcb280fd4eb027e1786e3766c7313f
1117b14
```

5. I captured the whole encrypted session traffic between the client and server. Now go to **Edit** | **Preferences** | **Protocol tree** | **SSL** | (Pre)-Master-Secret log filename. Enter `/path/to/sslkeylog.txt`and **OK**. Then right-click on the SSL packet (make sure you select **Decrypt packet data**. The option should be present in the bytes pane) and follow the SSL stream. Now you will see something like the following screenshot:

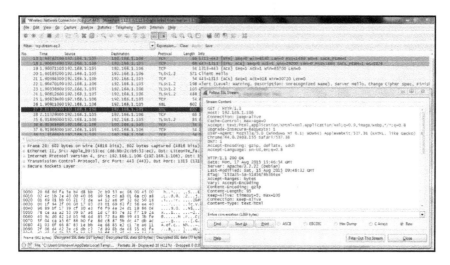

Decrypt SSL traffic

This is one of the easiest ways to decrypt SSL traffic with just a few clicks. One more way is to feed the RSA private key of the server into the Wireshark SSL preferences, which will give you the same result (I'm leaving it to you for your research).

Summary

DNS is a protocol used to resolve website names to an IP address. Through DNS, your machine is able communicate on an IP-based network.

FTP has been used to transfer files from one machine to another since the internet came into existence and is still being used in today's modern networks.

Web browsers present and transfer web-based content back and forth using HTTP. It is also commonly referred to as the request/response model, where a host requests a certain resource and the server responds with a status code and the resource if available.

SMTP is very commonly used to send emails. The SMTP command and its corresponding arguments are passed over the wire in plaintext.

VoIP traffic is made up of two things: RTP for data transfer and SIP for session creation. The signaling protocol creates and manages a session where RTP is used to carry the voice itself.

5
Analyzing the Transport Layer Protocols TCP/UDP

This chapter will help you understand the underlying technology enabling movement of network traffic across routing infrastructures through analysis of the transport layer protocols **Transmission Control Protocol** (**TCP**) and **User Datagram Protocol** (**UDP**). TCP and UDP are the basis of networking protocols and it is important to understand their structure and behavior.

The following are the topics that we will cover in this chapter:

- The TCP header and how it communicates
- Understanding the TCP flags
- Checking for different analysis flags in Wireshark
- Understanding UDP traffic
- Unusual patterns of TCP and UDP traffic

We will also look at some common anomalies that occur in day-to-day network operations.

The transmission control protocol

TCP is a connection-oriented protocol used by several application-layer protocols to ensure data delivery without any loss of information during transition, based on sequence and acknowledgment numbers. TCP ensures fail-proof delivery of packets between nodes. TCP sits in between the network layer and the application layer and uses the IP datagram to transfer data packets between the sender and receiver.

The **Three-Way Handshake** process takes place before the data transfer happens. A TCP connection is like a two-way communication process where not only the sender is actively involved, but even the receiver sends acknowledgments to make it a reliable form of connection.

Understanding the TCP header and its various flags

The TCP header is normally 20 bytes long, but at times, due to the presence of the `Options` field, the TCP header size can vary up to 60 bytes. The following is an illustration of a simplified TCP header:

Source port		Destination port	
Sequence number			
Acknowledgement number			
Data offset	Flags	Window size	
Checksum		Urgent pointer	
Options			

The following is a brief explanation for each of the TCP header fields:

- **Source port**: Used by the sending side to keep track of existing data streams and new incoming connections.
- **Destination port**: Port number associated with the services offered by the destination.
- **Sequence and acknowledgment numbers**: Each side uses a sequence number to keep track of ordering of the packets. Acknowledgment numbers are used by the sender and receiver to communicate the sequence number that is either received or sent.
- **Data offset:** Indicates where the data packet begins and the length of the TCP header. The size can vary due to the presence of the options field.
- **Flags**: There are various types of flag bits present; each of them has its own significance. They initiate connections, carry data, and tear down connections:

- **SYN (synchronize)**: Packets that are used to initiate a connection.
- **ACK (acknowledgment)**: Packets that are used to confirm that the data packets have been received, also used to confirm the initiation request and tear down requests
- **RST (reset)**: Signify the connection is down or maybe the service is not accepting the requests
- **FIN (finish)**: Indicate that the connection is being torn down. Both the sender and receiver send the FIN packets to gracefully terminate the connection
- **PSH (push)**: Indicate that the incoming data should be passed on directly to the application instead of getting buffered
- **URG (urgent)**: Indicate that the data that the packet is carrying should be processed immediately by the TCP stack
- **CWR (congestion window reduced)**: Used by either of the parties to slow down transmission speed in an event of congestion to avoid packet loss

- **Window size**: Indicates the amount of data that the sender can send. The size is decided during the handshake process to communicate and match the buffer size compatible for transmission.
- **Checksum**: Used by the receiving end to validate the integrity of the segments.
- **Urgent pointer**: Often marked as 0, used in conjunction with URG flag to mark immediate processing of a subset of message.
- **Options**: This field length can vary due to the presence of various options. This field has three parts: the first part specifies the length of the option field, the second part signifies the options being used, and the third contains the options in use. One of the important options, **maximum segment size** (**MSS**), is also part of this field.
- **Data**: The last part in the TCP header is the real data.

The preceding information gives us an overview regarding TCP headers and the significance of various parts of the header. While analyzing TCP sessions, it becomes quite important to know about these details.

How TCP communicates

To understand and analyze the packets in real time, I have configured a server that runs at 172.16.136.129 and a client that runs at 172.16.136.1, as shown in the following diagram:

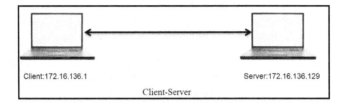

Client:172.16.136.1   Server:172.16.136.129

Client-Server

Using Wireshark, we will capture the three-way handshake process, which happens before the actual data transfer, as well as the teardown process (graceful termination).

How it works

The following screenshot depicts the various packets that are being generated while a client is trying to visit the web page hosted on http://172.16.136.129:

Use the following display filter to ease analysis:

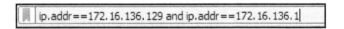

```
282 -895706969.756684000   172.16.136.1      172.16.136.129   TCP     64 52138→80 [SYN] Seq=0 Win=65535 Len=0
283 -1439969339.488273000  172.16.136.129    172.16.136.1     TCP     60 80→52138 [SYN, ACK] Seq=0 Ack=1 Win=2
284 15.671376000           172.16.136.1      172.16.136.129   TCP     52 52138→80 [ACK] Seq=1 Ack=1 Win=131744
285 15.672063000           172.16.136.1      172.16.136.129   HTTP    375 GET / HTTP/1.1
286 1228372207.391617000   172.16.136.129    172.16.136.1     TCP     52 80→52138 [ACK] Seq=1 Ack=324 Win=3072
287 15.672711000           172.16.136.129    172.16.136.1     HTTP    503 HTTP/1.1 200 OK  (text/html)
288 15.672725000           172.16.136.1      172.16.136.129   TCP     52 52138→80 [ACK] Seq=324 Ack=452 Win=13
289 -895706969.777480000   172.16.136.1      172.16.136.129   TCP     64 52139→80 [SYN] Seq=0 Win=65535 Len=0
290 15.747286000           172.16.136.129    172.16.136.1     TCP     60 80→52139 [SYN, ACK] Seq=0 Ack=1 Win=2
291 714245694.355758000    172.16.136.1      172.16.136.129   TCP     52 52139→80 [ACK] Seq=1 Ack=1 Win=131744
292 378319958.968279000    172.16.136.1      172.16.136.129   HTTP    359 GET /favicon.ico HTTP/1.1
293 1580695018.460033000   172.16.136.129    172.16.136.1     TCP     52 80→52139 [ACK] Seq=1 Ack=308 Win=3072
294 -459410977.038322000   172.16.136.129    172.16.136.1     HTTP    556 HTTP/1.1 404 Not Found  (text/html)
295 15.754902000           172.16.136.1      172.16.136.129   TCP     52 52139→80 [ACK] Seq=308 Ack=505 Win=13
299 20.679013000           172.16.136.129    172.16.136.1     TCP     52 80→52138 [FIN, ACK] Seq=452 Ack=324 W
300 609634608.344347000    172.16.136.1      172.16.136.129   TCP     52 52138→80 [ACK] Seq=324 Ack=453 Win=13
301 20.761722000           172.16.136.129    172.16.136.1     TCP     52 80→52139 [FIN, ACK] Seq=505 Ack=308 W
302 -1931345972.395708000  172.16.136.1      172.16.136.129   TCP     52 52139→80 [ACK] Seq=308 Ack=506 Win=13
```

A three-way handshake process is taking place in the packets 282, 283, and 284 to create a dedicated channel. The client initiated the creation by sending a SYN packet in the 282 packet with the SEQ set to 0. Since the server was open for communication, the server responded with a SYN/ACK packet with ACK set to 1 and SEQ set to 0, followed by a confirmation sent from the client side in the packet number 284 with SEQ=1 and ACK=1.

After the successful completion of channel creation, the client sends a GET request to access the contents of the web-root directory. The server acknowledges this in the packet number 287 and sends the requested content with the 200 OK status message, which is acknowledged by the client in the next packet.

After all the data transfer takes place, when the client has nothing left to request, or when the server has nothing left to send, the client sends FIN/ACK packets to properly terminate the connection. The server acknowledges this and sends its own FIN/ACK packets, which are acknowledged by the client in the packet number 302. This way of termination is often referred to as the teardown process. Refer to the following screenshot, which illustrates this process:

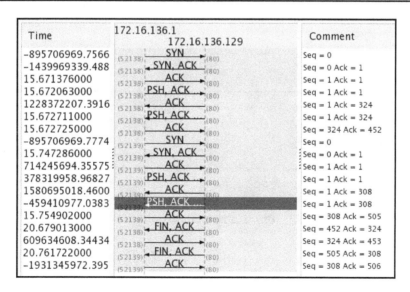

```
299 20.679013000          172.16.136.129   172.16.136.1     TCP   52 80-52138 [FIN, ACK] Seq=452 Ack=324
300 609634608.344347000   172.16.136.1     172.16.136.129   TCP   52 52138-80 [ACK] Seq=324 Ack=453 Win=1
301 20.761722000          172.16.136.129   172.16.136.1     TCP   52 80-52139 [FIN, ACK] Seq=505 Ack=308
302 -1931345972.395708000 172.16.136.1     172.16.136.129   TCP   52 52139-80 [ACK] Seq=308 Ack=506 Win=1
```

Time	172.16.136.1 / 172.16.136.129	Comment
-895706969.7566	(52138) SYN (80)	Seq = 0
-1439969339.488	(52138) SYN, ACK (80)	Seq = 0 Ack = 1
15.671376000	(52138) ACK (80)	Seq = 1 Ack = 1
15.672063000	(52138) PSH, ACK (80)	Seq = 1 Ack = 1
1228372207.3916	(52138) ACK (80)	Seq = 1 Ack = 324
15.672711000	(52138) PSH, ACK (80)	Seq = 1 Ack = 324
15.672725000	(52138) ACK (80)	Seq = 324 Ack = 452
-895706969.7774	(52139) SYN (80)	Seq = 0
15.747286000	(52139) SYN, ACK (80)	Seq = 0 Ack = 1
714245694.35575	(52139) ACK (80)	Seq = 1 Ack = 1
378319958.96827	(52139) PSH, ACK (80)	Seq = 1 Ack = 1
1580695018.4600	(52139) ACK (80)	Seq = 1 Ack = 308
-459410977.0383	(52139) PSH, ACK (80)	Seq = 1 Ack = 308
15.754902000	(52139) ACK (80)	Seq = 308 Ack = 505
20.679013000	(52138) FIN, ACK (80)	Seq = 452 Ack = 324
609634608.34434	(52138) ACK (80)	Seq = 324 Ack = 453
20.761722000	(52139) FIN, ACK (80)	Seq = 505 Ack = 308
-1931345972.395	(52139) ACK (80)	Seq = 308 Ack = 506

How sequence numbers are generated and managed

You must be wondering who assigns sequence number to packets and how. The device that initiates connection uses **Initial Sequence Numbers (ISN)** that are generated by the host's operating system. It can be any random number that has no significance with respect to the data. The sequence number we see in the packet one is zero is a relative referencing technique used by Wireshark.

Starting from packet 1, where SEQ=0 (the relative sequence number in real is 704809601), which is received by the server and in return replies with its own SEQ=0 and ACK=1 for the client's SEQ=0. At the end of this three-way handshake, the client replies with SEQ=1 and ACK=1 without any further increments as no data is being transferred during the process.

Then, by the fourth packet, the client sends a GET request with SEQ=1 and ACK=1 where the data payload length equals 323 (refer to the following screenshot), which the server receives and acknowledges with SEQ=1 and ACK=324. Did you see what just happened? The server replied by adding a total data payload length into ACK to denote that the data was successfully received:

```
▷ Frame 285: 375 bytes on wire (3000 bits), 375 bytes captured (3000 bits) on interface 0
▷ Raw packet data
▷ Internet Protocol Version 4, Src: 172.16.136.1 (172.16.136.1), Dst: 172.16.136.129 (172.16.136.129)
▽ Transmission Control Protocol, Src Port: 52138 (52138), Dst Port: 80 (80), Seq: 1, Ack: 1, Len: 323
     Source Port: 52138 (52138)
     Destination Port: 80 (80)
     [Stream index: 7]
     [TCP Segment Len: 323]
     Sequence number: 1     (relative sequence number)
     [Next sequence number: 324     (relative sequence number)]
```

RST (reset) packets

Often, there will be situations such as the server daemon is not available/running, the server is not able to process your request due to overload, you are restricted to interact with the server, or the port you are trying to connect to is not ready/open for connections. The RST packet basically denotes the abrupt rejection of a connection request.

In our scenario, the server daemon is not running and the client is trying to communicate; as a result, it receives RST packets in return for every SYN request sent. The client tries visiting the web page just once, but Wireshark captures more than one SYN and RST packet because every browser performs a different number of attempts over a non-responding or a closed socket at a preconfigured interval. Hence, in our case, I am using the Apple Safari browser, which made three attempts to connect in a span of 3-4 minutes. Refer to the following screenshot, which illustrates the packets captured in the process:

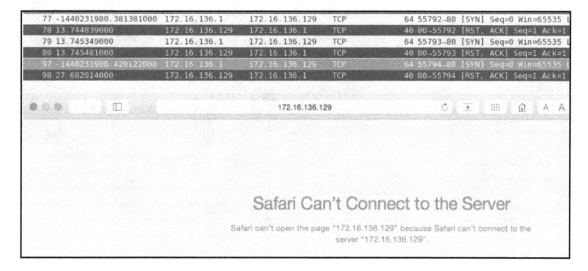

Unusual TCP traffic

Lost connection or unsuccessful connection attempt scenarios are the most common forms of unusual TCP traffic. You might also observe several other scenarios, such as high latencies due to long-distance communications. To make the analysis convenient and easy to troubleshoot, use the time column by sorting it to figure out large time gaps between the packets at the top of the list pane.

Another example can be where a malicious device is running a port scan on your network and your firewall responds with RST packets to avoid such reconnaissance attacks, or it might also be possible that the port closed. Refer to the following screenshot, where I've tried scanning a node over network using nmap, and it seems quite visible (due to a lot of packets generated from one source destined for random port numbers), and hence is easy to track:

17 42.896242000	172.16.136.129	172.16.136.1	TCP	44 52604→993 [SYN] Seq=1
18 -1440527712.212734000	172.16.136.1	172.16.136.129	TCP	40 993→52604 [RST, ACK]
19 42.896522000	172.16.136.129	172.16.136.1	TCP	44 52604→21 [SYN] Seq=18
20 42.896542000	172.16.136.1	172.16.136.129	TCP	40 21→52604 [RST, ACK] S
21 -1440526406.274558000	172.16.136.129	172.16.136.1	TCP	44 52604→113 [SYN] Seq=1
22 -1440529409.791742000	172.16.136.1	172.16.136.129	TCP	40 113→52604 [RST, ACK]
23 42.897040000	172.16.136.129	172.16.136.1	TCP	44 52604→554 [SYN] Seq=1
24 -1440529413.396222000	172.16.136.1	172.16.136.129	TCP	40 554→52604 [RST, ACK]
25 42.897314000	172.16.136.129	172.16.136.1	TCP	44 52604→143 [SYN] Seq=1
26 42.897326000	172.16.136.1	172.16.136.129	TCP	40 143→52604 [RST, ACK]
27 -1440527002.586622000	172.16.136.129	172.16.136.1	TCP	44 52604→111 [SYN] Seq=1
28 -1440529304.344318000	172.16.136.1	172.16.136.129	TCP	40 111→52604 [RST, ACK]
29 -1440529409.461758000	172.16.136.129	172.16.136.1	TCP	44 52604→256 [SYN] Seq=1
30 42.897884000	172.16.136.1	172.16.136.129	TCP	40 256→52604 [RST, ACK]
31 -1440529409.461758000	172.16.136.129	172.16.136.1	TCP	44 52604→8888 [SYN] Seq=
32 42.898151000	172.16.136.1	172.16.136.129	TCP	40 8888→52604 [RST, ACK]
33 -1440529409.461758000	172.16.136.129	172.16.136.1	TCP	44 52604→3389 [SYN] Seq=
34 42.898425000	172.16.136.1	172.16.136.129	TCP	40 3389→52604 [RST, ACK]
35 42.898743000	172.16.136.129	172.16.136.1	TCP	44 52604→23 [SYN] Seq=10

```
Frame 19: 44 bytes on wire (352 bits), 44 bytes captured (352 bits) on interface 0
Raw packet data
Internet Protocol Version 4, Src: 172.16.136.129 (172.16.136.129), Dst: 172.16.136.1 (172.16.136.1)
Transmission Control Protocol, Src Port: 52604 (52604), Dst Port: 21 (21), Seq: 1024978624, Len: 0
  Source Port: 52604 (52604)
```

Observe Frame 19, where the port scan initiated sents a SYN packet in order to check whether the port is open or closed. As a result, port 21 (FTP) was closed; hence the server sent an RST packet. There can be various scenarios other than the one discussed previously. If you hold a strong basic working knowledge of TCP and IP, then it would be quite easy for you to point out unusual forms of traffic.

The User Datagram Protocol

As defined in RFC 768, a UDP is a connectionless protocol, which is great for transmitting real-time data between hosts and is often termed as an unreliable form of communication. The reason is, UDP doesn't care about the delivery of packets, and any lost packets are not recovered because the sender is never informed about the dropped or discarded packets. However, many protocols such as DNS, TFTP, SIP, and so on. rely only on this.

The protocols that use UDP as a transport mechanism should rely upon other techniques to ensure data delivery and error-checking. A point to note is that UDP provides faster transmission of packets as it does not perform three-way handshake or graceful termination as observed in the TCP. UDP is referred to as a transaction-oriented protocol and not a message-oriented protocike a Tol lCP.

The UDP header

The size of a usual UDP header is 8 bytes; the data that is added with the header can be theoretically 65,535 (practically 65,507) bytes long. A UDP header is quite small when compared to a TCP header; it has just four common fields: **Source Port**, **Destination Port**, **Packet Length**, and **Checksum**. Refer to the UDP header shown here:

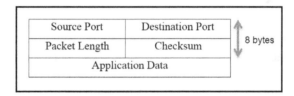

- **Source port**: Port number used by the sending side to receive any replies if needed. Most of the time, in a TCP and UDP, the port number chosen to be the part of the socket is ephemeral.
- **Destination port**: Port number used by the receiving side, where all data is transmitted to.
- **Packet length**: Specifies the length of the packet, starting from the header to the end of the data; the minimum length you will observe will be 8 bytes, that is the length of the UDP header.

- **Checksum**: Data integrity ensures that what is sent from the sender side is the same as what receiver got. Sometimes, while working with a UDP, you will see that the checksum value is 0 in the packet received. This means that the checksum is not required to be validated.

How it works

Let's analyze protocols such as DHCP, DNS, and TFTP, which use UDP as a delivery protocol.

I have configured a default gateway at 192.168.1.1 and a client at 192.168.1.106. Wireshark running between them will capture the UDP transactions. The following is a reference architecture diagram:

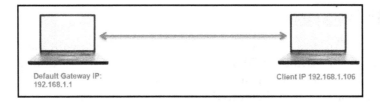

The DHCP

The protocol that manages IP addresses assigned to nodes and makes them network communication compatible is the **Dynamic Host Configuration Protocol** (**DHCP**). It is an automated way of assigning and managing IP addresses to requesting devices.

To generate DHCP packets from a client machine assigned with an IP address, I will try to release the current IP. Refer to the following screenshot:

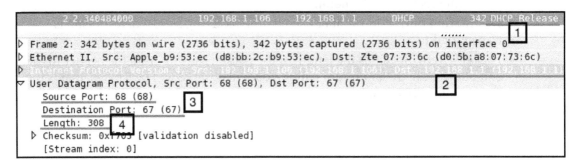

In the list pane, we can see a DHCP release packet that was sent explicitly by the client (I used the `dhclient -v -r` command on the Linux Terminal to release the IP address).

The DHCP server port number is `67` and the DHCP client port number is `68` by default. There is a fourth field that I have highlighted, the packet length field, which specifies the length of the packet, starting from the first byte until the end of data in the packet. However, out of 308 bytes, 8 bytes show the length of the UDP header and the remaining 300 bytes represent the application data.

The TFTP

The **Trivial File Transfer Protocol** (**TFTP**) is a lightweight version of the FTP that is used to transfer files between devices. Unlike the FTP protocol, TFTP does not ask users for any credentials. TFTP uses UDP as a transport mechanism.

Most commonly, TFTP is used in LAN environments and, when dealing with manageable devices such as switches and routers, network administrators use TFTP servers to take back up of configuration files and to update the firmware.

TFTP server is running at IP `192.168.1.106` and a TFTP client at IP `192.168.1.104`. There is a text file `abc.txt` stored on the TFTP server, which the TFTP client will download. Refer to the following diagram:

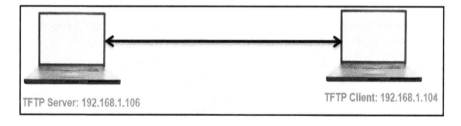

TFTP Server: 192.168.1.106   TFTP Client: 192.168.1.104

The traffic generated between two hosts is successfully captured and the packets corresponding to it are shown in the following screenshot

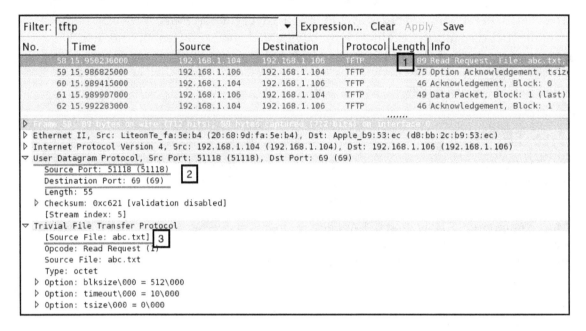

Now, let's see what each pointer signifies:

1. Depicts transfer of the packets is initiated as soon as the client requests the abc.txt file. The request frame can be seen in the list pane.

2. As discussed, a TFTP uses a UDP for a transport mechanism. The related details for the request are shown in the details pane, which states that the request was initiated from an ephemeral port number from the client destined to port 69 on the server (69 is a default port to the TFTP protocol).

3. The request was specific to the abc.txt file that is also present in the details pane in the TFTP protocol section.

Some applications use a UDP as a transport protocol and have their own built-in feature to ensure delivery. You must be wondering about the acknowledgment packets that are shared between the two hosts. As we discussed, a UDP is an unreliable form of communication, so why are we seeing ACKs in a UDP? The reason is that the TFTP server we are requesting has a built-in reliability feature.

Unusual UDP traffic

The following are a few traffic patterns that may be found suspicious in some environments.

Scenario 1: In a scenario where the UDP service is not running/available, what will the traffic look like then? Refer to the following screenshot:

The client requested an invalid resource that the server couldn't locate and hence returned with an error code and the summary message `File not found` (seen in the list pane).

Scenario 2: Sometimes, it is possible that the server daemon may not be running and the client may request a certain resource. In such cases, the client would receive the `ICMP destination unreachable` error with the error code `3`. Refer to the following screenshot:

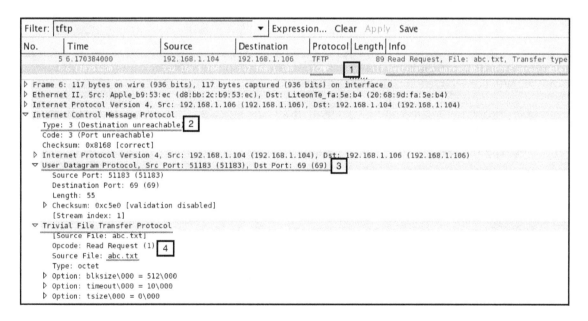

Let's discuss what each pointer signifies in more detail:

1. The server returned with an `ICMP destination unreachable` message when the TFTP server daemon was not functional
2. The client received an error code of type `3`
3. The request was sent to port `69`, which was currently nonfunctional
4. The requested resource shown under the TFTP protocol section

Scenario 3: Unusual DNS requests are also often seen when a client initiates a request to look for name servers associated with an address. It would look like the one shown in the following screenshot:

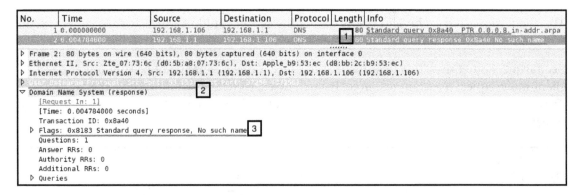

Now we will see what each pointer signifies:

1. As seen in the list pane, the client at `192.168.1.106` initiated a request to look for the address `8.0.0.0` and received a response in `Frame 2 No such Name`
2. The request was sent to the default gateway that holds the DNS cache
3. The gateway responded with a `No such name` error

There can be multiple scenarios where you will see unusual traffic related to UDP. Based on your usual network activity, it is advisable to create a traffic pattern to identify anomalies in DNS, DHCP, TFTP, and so on. UDP protocols.

 Learn about malicious DNS traffic to protect your digital infrastructure.

Summary

TCP is a reliable form of communication that facilitates three-way handshakes that and a teardown process ensures the connection is reliable and interactive.

A TCP header is 20 bytes long and consists of various fields such as source and destination port, SEQ and ACK numbers, offset, window size, flag bits, checksum, and options.

The SEQ and ACK numbers are used by TCP-based communications to keep track of data sent across.

A UDP is a connectionless protocol that is a nonreliable means of communication over IP, where the lost and discarded packets are never recovered. A UDP does provide faster transmission and easier creation of sessions.

A UDP header is 8 bytes long and has very few fields, such as source and destination port, packet length, and checksum. Common protocols such as DHCP, TFTP, DNS, and RTP mostly use a UDP as a transport mechanism.

6
Network Security Packet Analysis

Wireshark is an efficient utility packed with an advanced set of features that assist security professionals in performing passive analysis of network traffic to identify and point out malicious packets and anomalies.

This chapter will guide you through how to use Wireshark to analyze security issues, such as analyzing malware traffic and footprinting attempts. We will cover the following topics:

- Analyzing port scanning, footprinting, and attack/exploitation network traffic
- Dissecting malicious ARP traffic
- Analyzing brute force attacks
- Inspecting malicious traffic
- Creating display and capture filter signatures for malicious traffic

Using real-life scenarios simulated in a virtual network infrastructure, we will capture and understand malicious traffic patterns and replicate attacks such as information gathering and exploitation attempts. We will start from information gathering activity followed by an exploitation through a malicious `.exe` file. Then we will move on to understanding ARP poisoning traffic commonly used for performing **man-in-the-middle** (**MiTM**) attacks.

Information gathering

The probability and success factor of every attack depends on information gained through passive and active scanning of the network. Footprinting and reconnaissance are synonyms for the term *information gathering*.

The following diagram depicts the virtual/physical infrastructure we will be using for our analysis and for replicating the attacks:

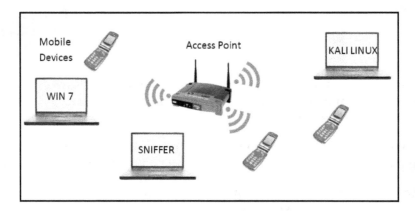

The access point is located at 192.168.1.1 and it allocates the IP address to connected devices using DHCP; the attacking box (Kali) is configured with a manual IP address 192.168.1.106.

PING sweep

Let's begin with our first scenario, where an attacker is trying to perform a ping sweep attack over the subnet his machine is a part of (assumption: The attacker is an internal employee). Refer to the following screenshot, which displays displays the traffic captured as a result of running a bash script (ping sweep scan); the script pings each IP, starting from 192.168.1.100 to 192.168.1.110:

No.	Time	Source	Destination	Protocol	Length	Info
1	0.000000000	Apple_b9:53:ec	Broadcast	ARP	42	Who has 192.168.1.110? Tell 192.168.1.106
2	0.004128000	Apple_b9:53:ec	Broadcast	ARP	42	Who has 192.168.1.109? Tell 192.168.1.106
3	0.008476000	Apple_b9:53:ec	Broadcast	ARP	42	Who has 192.168.1.108? Tell 192.168.1.106
4	0.012705000	Apple_b9:53:ec	Broadcast	ARP	42	Who has 192.168.1.107? Tell 192.168.1.106
5	0.023785000	192.168.1.106	192.168.1.105	ICMP	98	Echo (ping) request id=0x11a8, seq=1/256, ttl=64
6	0.027774000	192.168.1.104	192.168.1.106	ICMP	98	Echo (ping) reply id=0x11a3, seq=1/256, ttl=64
7	0.031652000	Apple_b9:53:ec	Broadcast	ARP	42	Who has 192.168.1.103? Tell 192.168.1.106
8	0.035462000	192.168.1.106	192.168.1.102	ICMP	98	Echo (ping) request id=0x1199, seq=1/256, ttl=64
9	0.040423000	192.168.1.106	192.168.1.101	ICMP	98	Echo (ping) request id=0x1194, seq=1/256, ttl=64
10	0.047374000	192.168.1.106	192.168.1.100	ICMP	98	Echo (ping) request id=0x118f, seq=1/256, ttl=64
11	0.122601000	LiteonTe_fa:5e:b4	Broadcast	ARP	42	Who has 192.168.1.106? Tell 192.168.1.105
12	0.124979000	Apple_b9:53:ec	LiteonTe_fa:5e:b4	ARP	42	192.168.1.106 is at d8:bb:2c:b9:53:ec
13	0.125118000	192.168.1.100	192.168.1.106	ICMP	98	Echo (ping) reply id=0x118f, seq=1/256, ttl=64
15	0.131304000	192.168.1.101	192.168.1.106	ICMP	98	Echo (ping) reply id=0x1194, seq=1/256, ttl=64
16	0.438404000	Apple_b9:53:ec	Zte_07:73:6c	ARP	42	Who has 192.168.1.1? Tell 192.168.1.106
17	0.528177000	Zte_07:73:6c	Apple_b9:53:ec	ARP	42	192.168.1.1 is at d0:5b:a8:07:73:6c

Ping sweep

Starting from packets 1-4, ARP requests are observed because of the ICMP `ping` command issued on Kali and, as it is fresh network, configuration devices would need to build their ARP cache table for internal LAN communication. In packet 5, the `ping` request is sent to `192.168.1.105`, and the reply for it is received in packet 14, which means the device is available. A similar pattern of traffic is captured and observed for the other IPs in the DHCP range. Due to frequent ARP and ICMP packets observed for a series of IPs one after another, we can conclude that it is a port scanning activity on the LAN network.

Half-open scan (SYN)

Now let's scan a specific device in the range of IP addresses and target the machine running at IP `192.168.1.105`. The primary way to gather information pertaining to a specific device would be a port scan in order to check for any open services that target device offers. By services, I mean HTTP daemons, mail server daemons, FTP server, SMB, and so on.

You might be wondering what a half-open scan is. Look at the process of a TCP three-way handshake we discussed in the previous chapter, where the client initiates the connection by sending a `SYN` packet and if the server is available client receives the `SYN`, `ACK` packet, and in return, the client sends an `ACK` packet to the server for completing the handshake process.

Now, what would happen if the `ACK` packet sent in the last step of the TCP handshake is never sent to the server? The server will wait for a period of time before terminating the handshake process, and the connection to the specific TCP service would never be completed. That's why this type of scan is called a half-open scan.

I have executed a half-open scan from the Kali box at IP `192.168.1.106` to target the Win7 box at IP `192.168.1.105` using Nmap with -sS switch. Nmap is an open source port scanning tool available for most platforms and can be downloaded for free from `http://nmap.org`. The traffic generated because of the `SYN` scan we executed is captured and shown in the following screenshot (use display filters for viewing packets pertaining to a specific host as follows):

Filter:	ip.addr==192.168.1.105		▼ Expression... Clear Apply Save			
No.	Time	Source	Destination	Protocol	Length	Info
13 0.312790000	192.168.1.106	192.168.1.105	TCP	58 34806-53 [SYN] Seq=1408496563 Win=0 MSS=1460
14 0.313002000	192.168.1.106	192.168.1.105	TCP	58 34806-1720 [SYN] Seq=1408496563 Win=1024 Len=0 MSS=1460
15 0.313161000	192.168.1.106	192.168.1.105	TCP	58 34806-1025 [SYN] Seq=1408496563 Win=1024 Len=0 MSS=1460
16 0.313362000	192.168.1.106	192.168.1.105	TCP	58 34806-3389 [SYN] Seq=1408496563 Win=1024 Len=0 MSS=1460
17 0.313502000	192.168.1.106	192.168.1.105	TCP	58 34806-23 [SYN] Seq=1408496563 Win=1024 Len=0 MSS=1460
18 0.313627000	192.168.1.106	192.168.1.105	TCP	58 34806-1723 [SYN] Seq=1408496563 Win=1024 Len=0 MSS=1460
19 0.313759000	192.168.1.106	192.168.1.105	TCP	58 34806-80 [SYN] Seq=1408496563 Win=1024 Len=0 MSS=1460
20 0.313886000	192.168.1.106	192.168.1.105	TCP	58 34806-993 [SYN] Seq=1408496563 Win=1024 Len=0 MSS=1460
21 0.314021000	192.168.1.106	192.168.1.105	TCP	58 34806-587 [SYN] Seq=1408496563 Win=1024 Len=0 MSS=1460
22 0.314148000	192.168.1.106	192.168.1.105	TCP	58 34806-113 [SYN] Seq=1408496563 Win=1024 Len=0 MSS=1460
25 0.410551000	192.168.1.105	192.168.1.106	TCP	54 113-34806 [RST, ACK] Seq=0 Ack=1408496564 Win=0 Len=0
26 0.413111000	192.168.1.106	192.168.1.105	TCP	58 34806-135 [SYN] Seq=1408496563 Win=1024 Len=0 MSS=1460
27 0.413276000	192.168.1.106	192.168.1.105	TCP	58 34806-554 [SYN] Seq=1408496563 Win=1024 Len=0 MSS=1460
28 0.416325000	192.168.1.105	192.168.1.106	TCP	58 135-34806 [SYN, ACK] Seq=2331129571 Ack=1408496564 Win=8
29 0.416892000	192.168.1.106	192.168.1.105	TCP	54 34806-135 [RST] Seq=1408496564 Win=0 Len=0
30 0.417633000	192.168.1.105	192.168.1.106	TCP	54 554-34806 [RST, ACK] Seq=0 Ack=1408496564 Win=0 Len=0
31 0.421378000	192.168.1.106	192.168.1.105	TCP	58 34806-443 [SYN] Seq=1408496563 Win=1024 Len=0 MSS=1460

Half-open scan

The key points/patterns to note in the above listed packets are as follows:

- There are numerous `SYN` packets generated from IP `192.168.1.106` destined for IP `192.168.1.105` over random ports within a very little amount of time. It is highly unlikely that an internal machine will initiate multiple connection instances within such short time frame (look at the time column).
- In the packets starting from 13 to 22, a `SYN` request is being sent so frequently over random and well-known port numbers within milliseconds.
- Also, the host at IP `192.168.1.106` never sent back a `ACK` packet in response to `SYN, ACK` received.

OS fingerprinting

Being aware of the operating system running on the target takes the information gathering process to the next level. If the make and version of operating system running is known to the attacker, it gives an extra edge in terms of exploitation through targeting specific vulnerabilities.

How do you think identifying the remote machine's OS works? I will tell you the secret. Every OS has a different way of implementing the TCP stack. So, a packet when received from the remote machine will have certain fields in it, such as TTL, fragment offset, and window size. By comparing the values in the packet with the database, tools are able to predict the OS with greater accuracy. For example, if you try to ping a Windows machine, the TTL value returned would be 128, and if you ping a Linux machine, the TTL value would be 64 most of the time. Simple, isn't it?

Using the nmap command `nmap -O 192.168.1.109,192.168.1.104`, let us fingerprint a machine's OS for IP `192.168.1.109` and `192.168.1.104` and capture the generated traffic.

We won't just rely on nmap's output to confirm the OS; we will also try to dissect packets from Wireshark for more clarity. Refer to the following screenshots to compare the outputs:

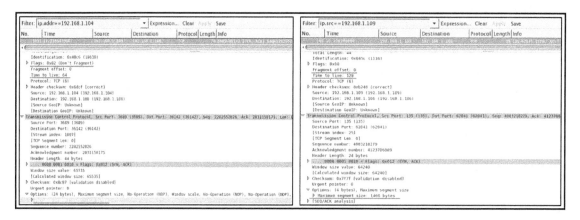

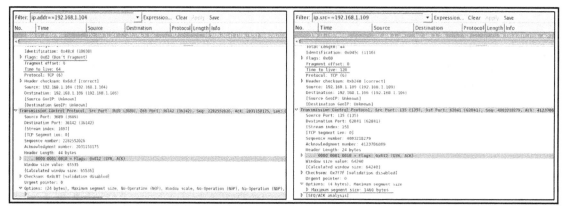

Check the highlighted TTL field value, which is equal to `64` for a Linux box and `128` for a Windows box. Also look at the maximum segment size value at the bottom where the value for a Linux box is `1460` and `1440` for a Windows box. Tools such as nmap store all these baseline values, which are then compared with scan results internally to identify the remote OS. A few key points to note to identify such malicious traffic are as follows:

- Traffic generated from the scans targeting to identify remote OS would be similar to the `SYN` scan (half-open) traffic, where the incomplete TCP handshakes and `ICMP` request/replies were observed.
- Also, if a lot of `RST` or `RST, ACK` packets are sent from a critical server to a specific host in a network, then it is something worth investigating further.

ARP poisoning

Whenever any device intends to communicate with another device, the requesting device sends a broadcast to the whole subnet. Then, the device to which the IP address belongs replies with its MAC address using a unicast packet. Through this approach, devices in local area network communicate with each other. A MAC address (physical address) table stores MAC address with its corresponding port number/IP address.

Use the `arp -a` command to populate the ARP table entries on your machine. The same command on a majority of platforms.

The following are some details pertaining to the local network we will be using for understating:

Device	IP address	MAC address
Router (default gateway)	192.168.1.1	D0:5B:A8:07:73:6C
Apple (victim)	192.168.1.103	D8:BB:2C:B9:53:EC
Windows server (victim)	192.168.1.109	00:0C:29:B3:CB:B6
Kali Linux (attacker)	192.168.1.106	00:0C:29:5D:A7:F7

For instance, if the Apple machine wishes to communicate with the Windows machine located at `192.168.1.109`, Apple will send a broadcast asking for the Windows MAC address stating `Who has 192.168.1.109? Tell 192.168.1.103`. Then, as soon as the Windows machine gets to know about the request, the ARP reply unicast packet stating `192.168.1.109 is at 00:0C:29:B3:CB:B6` will be sent.

ARP poisoning is an attack form to poison/infect/corrupt the local ARP cache of the victim. Refer to the following diagram:

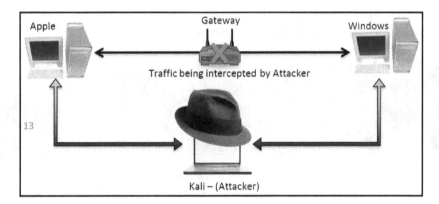

 IP forwarding is preconfigured using the command `echo '1'/proc/sys/net/ipv4/ip_forward` on a Kali box to send traffic back and forth between the Apple and Windows box.

Perform the following steps in order to replicate a MiTM attack in an lab environment:

1. The following screenshot shows the ARP table entry for both the client and server, before the attacker poisons the ARP cache for the victim machines:

```
Command Prompt                                                    _ □ ×

C:\Documents and Settings\Administrator>arp -a

Interface: 192.168.1.109 --- 0x10003
  Internet Address         Physical Address       Type
  192.168.1.103            d8-bb-2c-b9-53-ec      dynamic
  192.168.1.106            00-0c-29-5d-a7-f7      dynamic

C:\Documents and Settings\Administrator>_
```

Windows server cache

```
Anonymous:~ NotFound$ arp -a
? (172.16.136.1) at 0:50:56:c0:0:1 on vmnet1 ifscope permanent [ethernet]
? (172.16.158.1) at 0:50:56:c0:0:8 on vmnet8 ifscope permanent [ethernet]
? (192.168.1.1) at d0:5b:a8:7:73:6c on en1 ifscope [ethernet]
? (192.168.1.100) at f0:c1:f1:63:41:95 on en1 ifscope [ethernet]
? (192.168.1.106) at 0:c:29:5d:a7:f7 on en1 ifscope [ethernet]
? (192.168.1.109) at 0:c:29:b3:cb:b6 on en1 ifscope [ethernet]
```

Apple cache

2. The attacker is using the command-line utility **arpspoof** to poison the ARP entries through forged ARP reply packets:

```
root@kali:~/Desktop/        # arpspoof -i eth0 -t 192.168.1.109 192.168.1.103
0:c:29:5d:a7:f7 d8:bb:2c:b9:53:ec 0806 42: arp reply 192.168.1.103 is-at 0:c:29:5d:a7:f7
0:c:29:5d:a7:f7 d8:bb:2c:b9:53:ec 0806 42: arp reply 192.168.1.103 is-at 0:c:29:5d:a7:f7
0:c:29:5d:a7:f7 d8:bb:2c:b9:53:ec 0806 42: arp reply 192.168.1.103 is-at 0:c:29:5d:a7:f7
0:c:29:5d:a7:f7 d8:bb:2c:b9:53:ec 0806 42: arp reply 192.168.1.103 is-at 0:c:29:5d:a7:f7
0:c:29:5d:a7:f7 d8:bb:2c:b9:53:ec 0806 42: arp reply 192.168.1.103 is-at 0:c:29:5d:a7:f7
0:c:29:5d:a7:f7 d8:bb:2c:b9:53:ec 0806 42: arp reply 192.168.1.103 is-at 0:c:29:5d:a7:f7
```

ARP reply packets sent to the Windows server on behalf of the Apple device

```
root@kali:~/Desktop/        # arpspoof -i eth0 -t 192.168.1.103 192.168.1.109
0:c:29:5d:a7:f7 d8:bb:2c:b9:53:ec 0806 42: arp reply 192.168.1.109 is-at 0:c:29:5d:a7:f7
0:c:29:5d:a7:f7 d8:bb:2c:b9:53:ec 0806 42: arp reply 192.168.1.109 is-at 0:c:29:5d:a7:f7
0:c:29:5d:a7:f7 d8:bb:2c:b9:53:ec 0806 42: arp reply 192.168.1.109 is-at 0:c:29:5d:a7:f7
0:c:29:5d:a7:f7 d8:bb:2c:b9:53:ec 0806 42: arp reply 192.168.1.109 is-at 0:c:29:5d:a7:f7
0:c:29:5d:a7:f7 d8:bb:2c:b9:53:ec 0806 42: arp reply 192.168.1.109 is-at 0:c:29:5d:a7:f7
0:c:29:5d:a7:f7 d8:bb:2c:b9:53:ec 0806 42: arp reply 192.168.1.109 is-at 0:c:29:5d:a7:f7
0:c:29:5d:a7:f7 d8:bb:2c:b9:53:ec 0806 42: arp reply 192.168.1.109 is-at 0:c:29:5d:a7:f7
0:c:29:5d:a7:f7 d8:bb:2c:b9:53:ec 0806 42: arp reply 192.168.1.109 is-at 0:c:29:5d:a7:f7
```

ARP reply packets sent to Apple device on behalf of the Windows server

3. The traffic generated because of the preceding command looks like the following:

```
23 3.015821000 Vmware_5d:a7:f7    Vmware_b3:cb:b6    ARP    42 192.168.1.103 is at 00:0c:29:5d:a7:f7
24 5.016999000 Vmware_5d:a7:f7    Vmware_b3:cb:b6    ARP    42 192.168.1.103 is at 00:0c:29:5d:a7:f7

5 2.001262000 Vmware_5d:a7:f7    d8:bb:2c:b9:53:ec    ARP    42 192.168.1.109 is at 00:0c:29:5d:a7:f7
6 4.001992000 Vmware_5d:a7:f7    d8:bb:2c:b9:53:ec    ARP    42 192.168.1.109 is at 00:0c:29:5d:a7:f7
```

4. The packets sent from the Kali box forced the Apple and Windows machines to update their local ARP cache holding legit MAC addresses with the attacker's MAC address `00:0C:29:5D:A7:F7`:

Poisoned window's cache

```
Anonymous:~ NotFound$ arp -a
? (172.16.136.1) at 0:50:56:c0:0:1 on vmnet1 ifscope permanent [ethernet]
? (172.16.158.1) at 0:50:56:c0:0:8 on vmnet8 ifscope permanent [ethernet]
? (192.168.1.1) at d0:5b:a8:7:73:6c on en1 ifscope [ethernet]
? (192.168.1.100) at f0:c1:f1:63:41:95 on en1 ifscope [ethernet]
? (192.168.1.106) at 0:c:29:5d:a7:f7 on en1 ifscope [ethernet]
? (192.168.1.109) at 0:c:29:5d:a7:f7 on en1 ifscope [ethernet]
```

Poisoned Apple's cache

5. Now all the traffic sent between the Apple and Windows boxes will be forwarded through Kali. For verification purposes, I turned off the Windows server machine and tried sending ICMP packets from the Apple box:

```
Anonymous:~ NotFound$ ping 192.168.1.109
PING 192.168.1.109 (192.168.1.109): 56 data bytes
92 bytes from 192.168.1.106: Redirect Host(New addr: 192.168.1.109)
Vr HL TOS  Len   ID Flg  off TTL Pro  cks     Src      Dst
 4  5  00 0054 8554   0 0000  3f  01 7230 192.168.1.103  192.168.1.109
```

The preceding output ensures that the packets are being forwarded through `192.168.1.106`, hence making our ARP poisoning attack a success.

Create static ARP entries in critical machines to protect them from ARP spoofing attack; refer to the following screenshot for configuring static entries in a Windows box:

```
C:\Documents and Settings\Administrator>arp -s 192.168.1.103 d8-bb-2c-b9-53-ec

C:\Documents and Settings\Administrator>arp -a

Interface: 192.168.1.109 --- 0x10003
  Internet Address        Physical Address       Type
  192.168.1.103           d8-bb-2c-b9-53-ec      static
```

Adding a static entry to local ARP cache

Analysing brute force attacks

You must be aware of the popularity of brute force attacks. The chances of success are not very high, but also it is not impossible due to the lack of complex passwords configured in corporate machines. Brute force attack is a way to guess login passwords configured in devices using a tool that automates password guessing process.

To analyze malicious traffic of such nature, I will attempt to perform brute force over a preconfigured FTP service. FTP is used to transfer files efficiently with the assurance of integrity and confirmed delivery of the data in modern and critical network infrastructures.

For testing and our analysis purposes, I have configured one FTP server at 192.168.1.108 over a Windows 7 machine and the attacker is at IP 192.168.1.106 over a Kali machine.

Let's replicate and analyze the attack and normal FTP traffic pattern. Perform the following steps if you want to replicate it, but for educational purposes only:

1. Configure the FTP client and the FTP server using whatever platform suits your needs best and make sure the link between the FTP server and the client is working.
2. Now, first, we will try log in to the FTP server using a legitimate user and will record the traffic. Later, we will use the **Follow TCP stream** option in Wireshark to view the traffic details in easy-to-understand plain text format.
3. Refer to the following screenshot where I initiated the connection between from FTP the client. I then supplied the wrong credentials in the first attempt, and then used the correct ones in the second attempt:

```
                          ⟰ Charit — root@kali: ~ — ssh — 80×25
root@kali:~# nc -nv 192.168.1.108 21
(UNKNOWN) [192.168.1.108] 21 (ftp) open
220-FileZilla Server version 0.9.32 beta
220-written by Tim Kosse (Tim.Kosse@gmx.de)
220 Please visit http://sourceforge.net/projects/filezilla/
user charit
331 Password required for charit
pass abc
530 Login or password incorrect!
user charit
331 Password required for charit
pass charit
230 Logged on
help
214-The following commands are recognized:
   USER   PASS   QUIT   CWD    PWD    PORT   PASV   TYPE
   LIST   REST   CDUP   RETR   STOR   SIZE   DELE   RMD
   MKD    RNFR   RNTO   ABOR   SYST   NOOP   APPE   NLST
   MDTM   XPWD   XCUP   XMKD   XRMD   NOP    EPSV   EPRT
   AUTH   ADAT   PBSZ   PROT   FEAT   MODE   OPTS   HELP
   ALLO   MLST   MLSD   SITE   P@SW   STRU   CLNT   MFMT
214 Have a nice day.
quit
221 Goodbye
```

4. After I successfully logged in, I issued the `help` command to view a list of commands available followed by a **quit** to terminate the connection.

5. Wireshark captured the traffic between the FTP client and server; let's use the **follow TCP stream** option (right-click in **list pane** | **follow** | **TCP Stream**) to see the details:

```
Stream Content
220-FileZilla Server version 0.9.32 beta
220-written by Tim Kosse (Tim.Kosse@gmx.de)
220 Please visit http://sourceforge.net/projects/filezilla/
user charit
331 Password required for charit
pass abc
530 Login or password incorrect!
user charit
331 Password required for charit
pass charit
230 Logged on
help
214-The following commands are recognized:
   USER   PASS   QUIT   CWD    PWD    PORT   PASV   TYPE
   LIST   REST   CDUP   RETR   STOR   SIZE   DELE   RMD
   MKD    RNFR   RNTO   ABOR   SYST   NOOP   APPE   NLST
   MDTM   XPWD   XCUP   XMKD   XRMD   NOP    EPSV   EPRT
   AUTH   ADAT   PBSZ   PROT   FEAT   MODE   OPTS   HELP
   ALLO   MLST   MLSD   SITE   P@SW   STRU   CLNT   MFMT
214 Have a nice day.
quit
221 Goodbye
```

FTP assembled stream

6. Now, as we have analyzed the normal traffic patterns, let's see what would malicious FTP packets (such as the brute force attack attempts) would look like. I am `THC-hydra` to perform a brute force attack using a basic dictionary file.

7. Issue the `hydra -l <username> -P <password file> ftp://<you target's IP address>` command. Refer to the following screenshot:

```
root@kali:~# hydra -l charit -P pass.txt ftp://192.168.1.103
Hydra v7.6 (c)2013 by van Hauser/THC & David Maciejak - for legal purposes only

Hydra (http://www.thc.org/thc-hydra) starting at 2015-09-12 18:16:00
[DATA] 11 tasks, 1 server, 11 login tries (l:1/p:11), ~1 try per task
[DATA] attacking service ftp on port 21
[21][ftp] host: 192.168.1.103   login: charit   password: charit
1 of 1 target successfully completed, 1 valid password found
Hydra (http://www.thc.org/thc-hydra) finished at 2015-09-12 18:16:04
```

8. The traffic generated was captured and, instead of displaying all the traffic, I have used a display filter `ftp.request.command==PASS` in order to view only packets pertaining to the FTP password command. The following screenshot shows what display filter I used to query malicious repetitive packets.

Filter: ftp.request.command == "PASS" ▼ Expression... Clear Apply Save						
No.	Time	Source	Destination	Protocol	Length	Info
60	1.169458000	192.168.1.106	192.168.1.103	FTP	76	Request: PASS 007
61	1.169645000	192.168.1.106	192.168.1.103	FTP	76	Request: PASS mno
62	1.169830000	192.168.1.106	192.168.1.103	FTP	79	Request: PASS charit
63	1.170013000	192.168.1.106	192.168.1.103	FTP	77	Request: PASS root
128	3.500600000	192.168.1.106	192.168.1.103	FTP	76	Request: PASS 123
131	3.501315000	192.168.1.106	192.168.1.103	FTP	76	Request: PASS efg
132	3.501529000	192.168.1.106	192.168.1.103	FTP	76	Request: PASS abc
133	3.502078000	192.168.1.106	192.168.1.103	FTP	78	Request: PASS admin
134	3.502479000	192.168.1.106	192.168.1.103	FTP	78	Request: PASS chris
136	3.503548000	192.168.1.106	192.168.1.103	FTP	76	Request: PASS mno

FTP Brute Force attack traffic pattern

9. It is easily identifiable that the preceding traffic is malicious due to the FTP pass command issued by a single IP over a very short period (refer to the time column).

To identify such malicious or sensitive traffic, create a different coloring scheme (discussed in Chapter 3, *Analysing Transport Layer Protocols TCP/UDP*). Refer to the following screenshot:

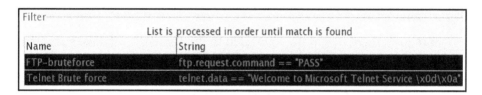

Coloring scheme for malicious traffic

Using a display filter and colorzing traffic option, you can analyze such malicious traffic in a network infrastructure.

Inspecting malicious traffic (malware)

Malware is one of the most common forms of client-side attacks in any network. The outcome of malware infections can be very damaging, ranging from denial of service attacks to remote code execution. Critical infrastructure industries such as Oil and Gas, Energy, Transport, and Manufacturing are one of the favorite targets for malware due to a lack of security controls and general awareness in place. Refer to the following screenshot, where we will try to replicate a malware-based infection in a lab:

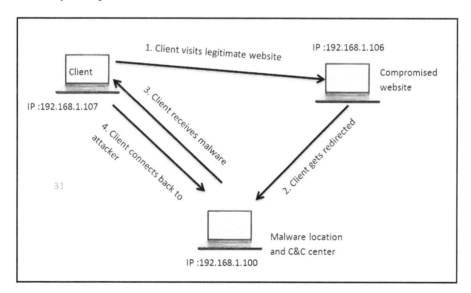

Malware is capable of performing tasks once installed on the victim's machine, such as information disclosure, executing commands, and/or corrupting systems, even if the best security solutions are installed in the infrastructure.

Follow these steps if you want to replicate the scenario in your own virtual lab:

- You require three machines connected to the same LAN. Make sure they are able to ping to each other, to ensure connectivity.
- On the IP address 192.168.1.106 stays a legitimate website, which the client at IP 192.168.1.107 usually visits. However, this time, the client is not aware of the infection that causes redirection to another web server (assumption: the web server is compromised and taken over by the attacker). Refer to the following screenshot of the legitimate server:

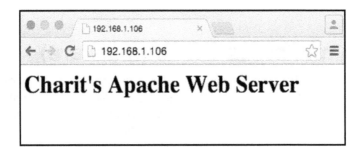

Legitimate website

- To simulate the redirection, I have configured my Apache server running on 192.168.1.106 to redirect HTTP requests to IP 192.168.1.100 and download the efg.exe from there.
- When client visits the website running at 192.168.1.106, it gets redirected to a new web server, which directly asks the client to run a file named efg.exe. Refer to the following screenshot:

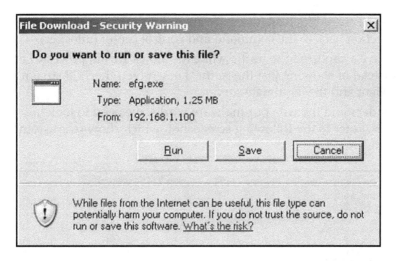

Client gets redirected to IP 192.168.1.100 and is asked to run the application.

- The publisher of the application is not verified, so the client operating system is not able to verify it. This results in an unknown publisher error. Refer to the following screenshot:

Unknown publisher error

- Once the client hits **Run**, the malware will be executed, thus creating a connection back to the command and control center (attacker).
- We have a captured the traffic while the attack was in process. Let's take a look at it. Instead of showing just the traffic, I assembled the TCP stream first between the client and the legitimate server.
- To understand the way our malware works, we need to look into the packet details. Refer to the following screenshot, which shows the assembled TCP stream:

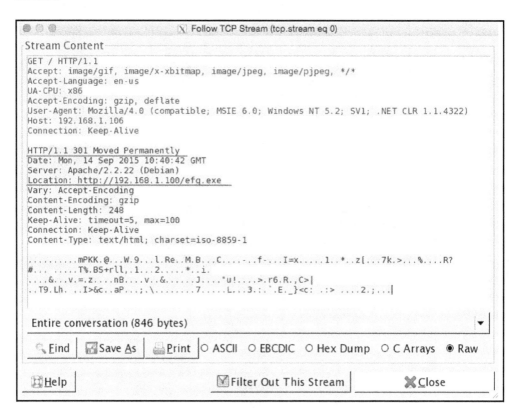

TCP stream between the client and real (compromised) server

- As is clearly visible, the client visits the web server, and the request is being forwarded with HTTP redirection to the new address `http://192.168.1.100/efg.exe`

- After a couple of packets were exchanged between the client and server, the client received a 200 OK status message, suggesting successful download of the executable application efg.exe

1255 36.428063(192.168.1.100   192.168.1.107   HTTP   1458 HTTP/1.1 200 OK (application/x-msdownload)

The following screenshot depicts the request sent by the client machine to download the executable from the new web address:

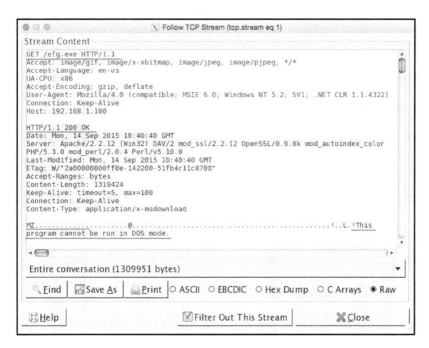

Figure 7.20: Malware signature

The GET request was initiated by the client in search of efg.exe, to which the server responded with a 200 OK status message. Later, you can see the known malware signature starting with the characters MZ followed by some random character.

A quick Google search reveal that it is an executable file. Wikipedia states 16/32 bit DOS executable files can be identified by the letters MZ at the beginning of the file in ASCII. Refer to the following screenshot:

DOS [edit]

Main articles: DOS MZ executable and New Executable

16-bit DOS MZ executable

The original DOS executable file format. These can be identified by the letters "MZ" at the beginning of the file in ASCII.

Moving on with our investigation, let's export the efg.exe file. Perform the following steps to download the file:

1. Go to **File** | **Export Objects** | **HTTP**:

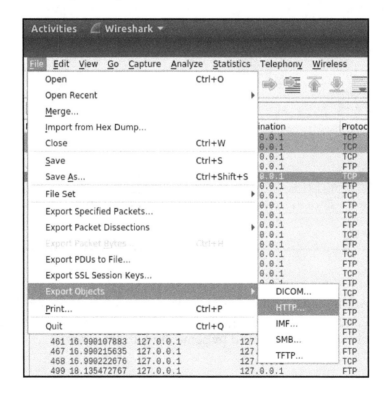

The next screen would look as the following screenshot:

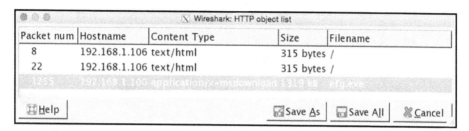

Packet num	Hostname	Content Type	Size	Filename
8	192.168.1.106	text/html	315 bytes /	
22	192.168.1.106	text/html	315 bytes /	
1255	192.168.1.100	application/x-msdownload	1319 kB	efg.exe

Exporting HTTP objects

2. Now, select the conversation that states the name of the file along with it and click on **Save As**.

3. An option is to upload this file to websites such as `http://www.virustotal.com`, which will scan the file through multiple antivirus programs. Refer to the following screenshot:

Uploadingefg.exeto virustotal.com

4. **Click Scan** and wait for the results:

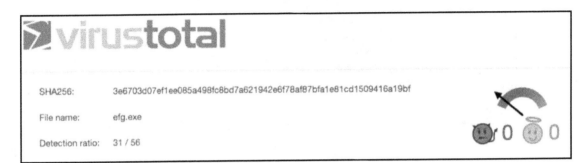

SHA256: 3e6703d07ef1ee085a498fc8bd7a621942e6f78af87bfa1e81cd1509416a19bf

File name: efg.exe

Detection ratio: 31 / 56

31 out of 56 type of antivirus software detected the executable file as malicious.

5. You can also manually examine the conversation between the infected client and the command and control center by looking at the `hex` dump. Refer to the following screenshot:

```
Stream Content
000A1978  46 69 6c 65 54 69 6d 65   54 6f 4c 6f 63 61 6c 46   FileTime ToLocalF
000A1988  69 6c 65 54 69 6d 65 00   ec 01 47 65 74 46 69 6c   ileTime. ..GetFil
000A1998  65 49 6e 66 6f 72 6d 61   74 69 6f 6e 42 79 48 61   eInforma tionByHa
000A19A8  6e 64 6c 65 00 00 8d 03   50 65 65 6b 4e 61 6d 65   ndle.... PeekName
000A19B8  64 50 69 70 65 00 fb 01   47 65 74 46 75 6c 6c 50   dPipe... GetFullP
000A19C8  61 74 68 4e 61 6d 65 57   00 00 bf 01 47 65 74 43   athNameW ....GetC
000A19D8  75 72 72 65 6e 74 44 69   72 65 63 74 6f 72 79 57   urrentDi rectoryW
000A19E8  00 00 d4 02 48 65 61 70   53 69 7a 65 00 00 53 04   ....Heap Size..S.
000A19F8  53 65 74 45 6e 64 4f 66   46 69 6c 65 00 00 73 01   SetEndOf File..s.
000A1A08  49 6d 70 65 72 73 6f 6e   61 74 65 4c 6f 67 67 65   Imperson ateLogge
000A1A18  64 4f 6e 55 73 65 72 00   1f 00 41 64 6a 75 73 74   dOnUser. ..Adjust
000A1A28  54 6f 6b 65 6e 50 72 69   76 69 6c 65 67 65 73 00   TokenPri vileges.
000A1A38  96 01 4c 6f 6f 6b 75 70   50 72 69 76 69 6c 65 67   ..Lookup Privileg
000A1A48  65 56 61 6c 75 65 41 00   00 00 00 00 00 00 00 00   eValueA. ........
000A1A58  00 00 00 00 00 00 00 00   00 00 00 00 00 00 00 00   ........ ........
```

Hexdump in TCP stream dialog

It seems that the attacker machine is issuing some command to gather information regarding the victim machine. The highlighted content on the right-hand side of the window states strings such as `Get File Information`, `Get full PC name`, `Get Current directory`, `Adjust token Privileges`, and so on.

Familiarity with such traffic patterns is critical, and it is advisable to set up filters capture filters in Wireshark to perform passive analysis to identify malicious traffic. For sure, IDS/IPS systems in your environment would be able to detect it automatically but in critical infrastructure networks (Oil and Gas, Energy, and so on), it is highly unlikely to have such security solutions deployed. In those scenarios, Wireshark is your best buddy and, most importantly, it comes for free!!

Summary

Use Wireshark to keep your network secure by defending against common forms of infiltration attempts. Analyzing the packets from a security perspective will give you a new insight into how to deal with malicious users.

Activities such as port scanning, footprinting, and various active information gathering attempts are the basis of attacking methodologies that can be taken advantage of to bypass your security infrastructure. Create filters and signatures to identify malicious traffic patterns.

Guessing passwords to gain unauthorized access is called a brute force attack. Through Wireshark, you can filter and identify such malicious forms of traffic.

Wireshark can help you in analyzing malware behavior, and using the behavior analyzed, you would be able to create the necessary signatures for your IDS/IPS security solutions.

The next chapter will enable network professionals to perform wireless packet analysis and teach them how to decrypt and read traffic from the air.

7
Analyzing Traffic in Thin Air

Most devices today are installed with wireless capabilities and it is critical to understand the structure and pattern of wireless traffic within your network. This chapter will assist in understanding the methodology and steps involved in performing wireless packet analysis.

The following are the topics we will cover in this chapter:

- Understanding IEEE 802.11
- Modes in wireless communication
- Capturing wireless traffic
- Analyzing normal and unusual traffic patterns
- Decrypting encrypted wireless traffic

Wireless network traffic analysis is similar to wired network analysis; the objective of the topics discussed here is to learn about wireless technologies and protocol strengths and weaknesses, along with suspicious wireless traffic.

Understanding IEEE 802.11

At the **Institute of Electrical and Electronics Engineers (IEEE)**, several groups of technical professionals as a committee are working on projects, and one of these is 802, which is responsible for developing **Local Area Networks (LAN)** standards. Specifically, 802.11 contains WLAN standards.

There are a couple of 802.11 standards, for an outmost coverage of standards we will discuss the multiple of them such as 802.11b, 802.11a, 802.11g, and 802.11n:

- **802.11**: Supports a network bandwidth of 1-2 Mbps. This is the reason why many 802.11-compatible devices have become obsolete.
- **802.11b**: This specification uses a signaling frequency of 2.4 Ghz like the 802.11 standard. Technically, a maximum of 11 Mbit transmission rate can be achieved over a 2.4 Ghz band using b specification.

The 802.11b band is divided into 14 overlapping channels, where every channel has 22 Mhz widths. In one instance, there can be a maximum of three non-overlapping channels operating at the same time. This space separation is necessary and required to let the channels operate individually.

Most appliances, such as microwave, cordless phones, and so on. work over a 2.4 Ghz spectrum, which may cause significant interference and congestion in 802.11b WLAN packets transmission.

- **802.11a**: This is based on **Orthogonal Frequency Division Multiplexing (OFDM)** that was released in 1999 and supports a maximum transmission rate up to 54 Mbps 5 Ghz spectrums. This specification was developed as a second standard to 802.11 standards. It is commonly used in business environments; because of its high cost, the *a* specification is not best suited for home environments. There is no channel overlap that happens in 802.11a. A higher regulated frequency helps in preventing the interferences caused by devices that work on 2.4 Ghz spectrums.
- **802.11g**: Released in 2002, this specification combines the best features of 802.11a and 802.11b. It uses a signaling frequency of 2.4 Ghz, and bandwidth up to 54 Mbps. It also supports backward compatibility, which means that all 802.11g access points will support network adapters using 802.11b and vice versa.

- **802.11n**: To improve further on the range and the transfer rates, wireless specification *n* was introduced based on technology **Multiple-Input Multiple-output** (**MIMO**). The final version of this specification, released in 2007, stated a transfer rate up to 600 Mbps. It can be configured with 2.4 or 5 Ghz; it can use both frequencies at the same time, thus enabling backward compatibility with network adapters. A maximum of four antennas can be used with the MIMO technology.

Various modes in wireless communications

Wireless networks uses the **Carrier Sense Multiple Access and Collision Avoidance** (**CSMA/CA**) protocol to manage the stations sending data, where every host that wants to send data is supposed to listen to the channel first, that is, if it is free, then the host can go ahead and send the packet; if not, then the host has to wait for its turn. This is because the same medium is being shared by every host, thus avoiding collisions.

The 802.11 architecture is composed of several components such as a **station** (**STA**), a wireless **Access Point** (**AP**), **Basic Service Set** (**BSS**), **Extended Service Set** (**ESS**), **Independent Basic service set** (**IBSS**), and **Distribution System** (**DS**).

There are four common modes of association between the STA and the AP, which are as follows:

- **Infrastructure/managed mode:** A wireless network where a wireless client establishes a connection with an access point to access data and network resources. An AP is defined with a **Service Set Identifier** (**SSID**), which is the access point's name used for identification purposes within a certain range (for security reasons, sometimes, broadcasting an SSID can be disabled, which will prevent your wireless network from being discovered). For example, when we scan for available nearby Wi-Fi networks around, we will be shown multiple network names to choose from. Another useful term to know is **Base Service Set Identifier** (**BSSID**), that is, the access point's MAC address.

By default, every access point is supposed to broadcast the SSID and transmit a beacon frame 10 times in a second to let clients know that AP is ready to accept connections. Refer to the following diagram:

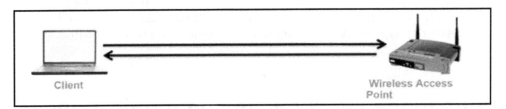

- **Ad Hoc mode**: In Ad Hoc mode, a peer-to-peer network is formed where two clients connect to each other. The packets sent and received by the wireless clients are not relayed to the access point. Refer to the following diagram:

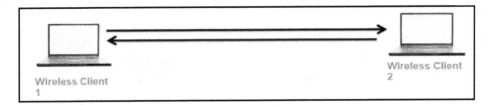

- **Master mode**: When the NIC (network interface card) adapter is capable to act as an access point for clients through usage of special drivers then it becomes a master node. Modern operating systems and hardware are enabled with such a feature, where the host device can act as an access point by sharing its wired connection. Refer to the following diagram:

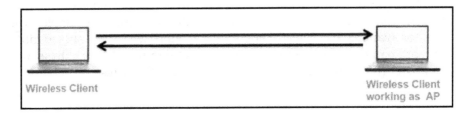

- **Monitor mode**: This mode enables a network adapter to listen to wireless network traffic; when the monitor mode is activated, your device will stop transmitting and receiving any packets and it will just sniff live traffic in a passive way. In short, wireshark running with an interface enabled with monitor mode can sniff traffic without being a part of the network. This mode is often termed as the **Radio Frequency Monitor Mode (RFMON)**. Refer to the following diagram:

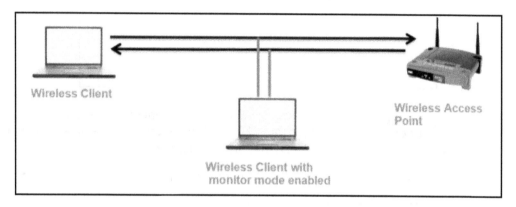

Usual and unusual wireless traffic

In 2003, **Wi-Fi Protected Access (WPA)** was launched by Wi-Fi Alliance as a measure to make WLAN communication stronger than the previous protocol, WEP. The key size used by WEP is 40/104 bits, whereas WPA uses a key size of 256 bits and also facilitates integrity checks. In WEP, the traditional CRC was implemented, but WPA introduced, the popular Michael 64-bit **Message integrity check (MIC)**.

WPA uses the RC4 algorithm to build a session based on dynamic encryption keys (you would never end up using the same key pair between two hosts). Refer to the following illustration of how the cipher text is formed that is transmitted over the medium:

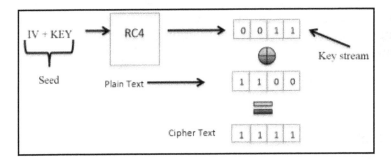

The process starts by appending the IV and the dynamically generated 256-bit key. Followed by encryption using RC4 algorithm, the resulting encrypted key stream is then appended with the data and voila! We have the final cipher text.

Refer to the following diagram depicting the authentication process in WPA:

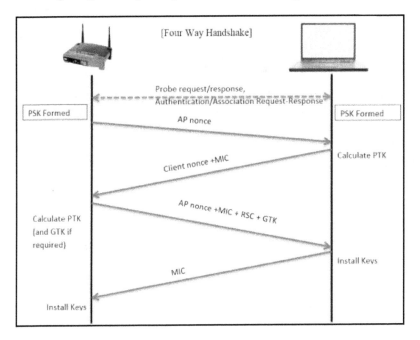

The following is a summary of steps involved for the preceding diagram:

1. First, the *Master Key Exchange (PSK)* takes place, followed by transmission of a nonce value to STA (initiation of connection).

2. The STA will use the AP's nonce value and its own nonce to calculate the **Pairwise Transient Key** (PTK) to send along with the **Pre-Shared Key** (PSK) established in the previous step. The resulting value will be sent to the AP to calculate the PTK and append the MIC with the **Receive Sequence Counter** (RSC).

3. Now, the STA will first verify the MIC in the message to ensure the integrity and then install the keys.

4. A response will be sent to the AP regarding the status. If the status shows success, the AP then installs the same keys (dynamic keys) that will be used in further communication.

The following screenshot depicts the four authentication packets involved in a successful WPA Enterprise handshake process:

```
Filter: eapol                                    ▼ | Expression... Clear  Apply  Save

No.    | Time           | Source         | Destination    | Protocol | Length | Info
257 8.730625000          Zte_07:73:6c     Apple_b9:53:ec   EAPOL              173 Key (Message 1 of 4)
259 8.733391000          Apple_b9:53:ec   Zte_07:73:6c     EAPOL              197 Key (Message 2 of 4)
265 8.736180000          Zte_07:73:6c     Apple_b9:53:ec   EAPOL              203 Key (Message 3 of 4)
267 8.737817000          Apple_b9:53:ec   Zte_07:73:6c     EAPOL              173 Key (Message 2 of 4)

▷ Frame 257: 173 bytes on wire (1384 bits), 173 bytes captured (1384 bits) on interface 0
▷ Radiotap Header v0, Length 36
▷ IEEE 802.11 QoS Data, Flags: ......F.C
▷
▽ 802.1X Authentication
    Version: 802.1X-2001 (1)
    Type: Key (3)
    Length: 95
    Key Descriptor Type: EAPOL WPA Key (254)
  ▷ Key Information: 0x008a
    Key Length: 16
    Replay Counter: 0
    WPA Key Nonce: 5ec313cec318318d18df8dffdffb0047fb8a47518aea5152...
    Key IV: 00000000000000000000000000000000
    WPA Key RSC: 0000000000000000
    WPA Key ID: 0000000000000000
    WPA Key MIC: 00000000000000000000000000000000
    WPA Key Data Length: 0
```

Getting into more detail, let's observe the flags involved in all of the preceding four authentication packets in the handshake process:

```
▽ 802.1X Authentication                           ▽ 802.1X Authentication
    Version: 802.1X-2001 (1)                          Version: 802.1X-2001 (1)
    Type: Key (3)              Packet 1               Type: Key (3)              Packet 2
    Length: 95                                        Length: 119
    Key Descriptor Type: EAPOL WPA Key (254)          Key Descriptor Type: EAPOL WPA Key (254)
  ▽ Key Information: 0x008a                          ▽ Key Information: 0x010a
    .... .... .... .010 = Key Descriptor Version: AES   .... .... .... .010 = Key Descriptor Version: AES
    .... .... .... 1... = Key Type: Pairwise Key        .... .... .... 1... = Key Type: Pairwise Key
    .... .... ..00 .... = Key Index: 0                  .... .... ..00 .... = Key Index: 0
    .... .... .0.. .... = Install: Not set              .... .... .0.. .... = Install: Not set
    .... .... 1... .... = Key ACK: Set                  .... .... 0... .... = Key ACK: Not set
    .... ...0 .... .... = Key MIC: Not set              .... ...1 .... .... = Key MIC: Set
    .... ..0. .... .... = Secure: Not set               .... ..0. .... .... = Secure: Not set
    .... .0.. .... .... = Error: Not set                .... .0.. .... .... = Error: Not set
    .... 0... .... .... = Request: Not set              .... 0... .... .... = Request: Not set
    ...0 .... .... .... = Encrypted Key Data: Not set   ...0 .... .... .... = Encrypted Key Data: Not set
    ..0. .... .... .... = SMK Message: Not set          ..0 .... .... .... = SMK Message: Not set
▽ 802.1X Authentication                           ▽ 802.1X Authentication
    Version: 802.1X-2001 (1)                          Version: 802.1X-2001 (1)
    Type: Key (3)              Packet 3               Type: Key (3)              Packet 4
    Length: 125                                       Length: 95
    Key Descriptor Type: EAPOL WPA Key (254)          Key Descriptor Type: EAPOL WPA Key (254)
  ▽ Key Information: 0x01ca                          ▽ Key Information: 0x010a
    .... .... .... .010 = Key Descriptor Version: AES   .... .... .... .010 = Key Descriptor Version: AES
    .... .... .... 1... = Key Type: Pairwise Key        .... .... .... 1... = Key Type: Pairwise Key
    .... .... ..00 .... = Key Index: 0                  .... .... ..00 .... = Key Index: 0
    .... .... .1.. .... = Install: Set                  .... .... .0.. .... = Install: Not set
    .... .... 1... .... = Key ACK: Set                  .... .... 0... .... = Key ACK: Not set
    .... ...1 .... .... = Key MIC: Set                  .... ...1 .... .... = Key MIC: Set
    .... ..0. .... .... = Secure: Not set               .... ..0. .... .... = Secure: Not set
    .... .0.. .... .... = Error: Not set                .... .0.. .... .... = Error: Not set
    .... 0... .... .... = Request: Not set              .... 0... .... .... = Request: Not set
    ...0 .... .... .... = Encrypted Key Data: Not set   ...0 .... .... .... = Encrypted Key Data: Not set
    ..0 .... .... .... = SMK Message: Not set           ..0 .... .... .... = SMK Message: Not set
```

Here is the description of the preceding authentication packets:

- **Packet 1**: The pairwise master key (pre-shared key) and the ACK bit are set (because of the association request/response exchanged earlier), which is sent by the AP to the STA to initiate the connection along with the nonce value chosen randomly.
- **Packet 2**: The pairwise master key and the MIC flag is set, which the STA sent to the AP to acknowledge the request, along with its own nonce value appended to the AP's nonce and the MIC for integrity check.

- **Packet 3**: The pairwise master key, install, key `ACK`, and `MIC` flags are set, which AP sent to STA. Next, the STA will fulfill the challenge in order to get authenticated.
- **Packet 4**: The pairwise master key and the `MIC` flag are set, which the STA sends to AP to complete the connection process.

Based on our understanding of a successful authentication process, now let's try to observe the packet pattern in the case of unsuccessful authentication. The only difference in this scenario is that the STA is not aware of the pre-shared key.

Refer to the following screenshot:

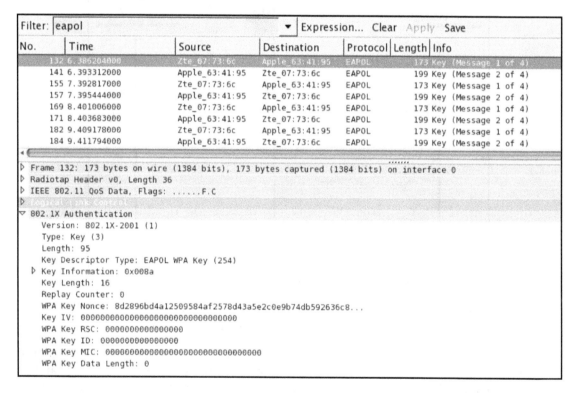

WPA Failed authentication

The preceding screenshot depicts the chain of packets transmitted between the STA and AP, due to an incorrect pre-shared key being sent by STA. These packets may be witnessed in an event of a brute force attack against the AP.

WPA Enterprise

In order to standardize and harden the authentication process and introduce a few elements of accountability and non-repudiation, WPA offers the configuration of an external server to validate and authorize STAs. This centralized authentication and validation unit is termed as a **RADIUS/TACACS** server. Before the four-way handshake takes place, the **RADIUS** server and the access point are supposed to go through a MSK. Let's have a look at the following diagram:

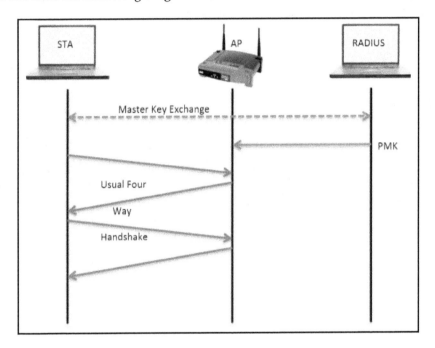

Post the exchange of the master key, the pairwise master key is created and passed on to the AP, which will further complete the four-way handshake process.

As a part of graceful termination, the wireless stations use disassociation packets in order to notify the access point that the STA is now going offline and the resources allocated can be released. The following screenshot lists the packets observed during the disassociation phase:

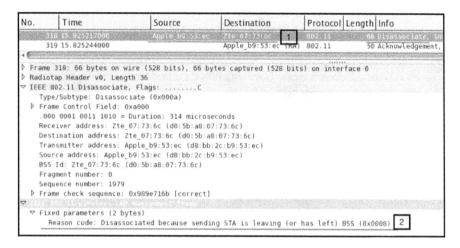

The disassociation packet

The wireless stations use the `deauthentication` frames to notify the access point that the STA is leaving. As we can observe in the preceding screenshot, first, the STA sends a `disassociation` frame and receives ACK (318,319) from AP.

There can be several scenarios where an wireless client would send a `disassociation` frame. Refer to the following screenshot to understand this:

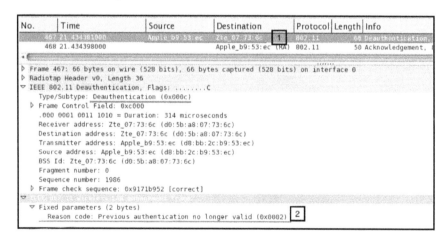

The deauthentication packet

In the preceding list of packets, first, the STA sends a `deauthentication` frame to the access point, which gets acknowledged in the next packets (467,468).

Decrypting wireless network traffic

Wireshark also facilitates decryption of wireless traffic through embedding a pre-shared key under the 802.11 protocol section. The following screenshot depicts normal wireless traffic being sniffed from a nearby access point:

No.	Time	Source	Destination	Protocol	Length	Info
2	0.000004	Tp-LinkT_2a:84:4e	MS-NLB-PhysServer-10_	802.11	145	QoS Data, SN=197, FN=0, Flags=.p....F.
3	0.101892	MS-NLB-PhysServer-10_	Tp-LinkT_2a:84:4e	802.11	26	QoS Null function (No data), SN=2641, FN=0, Flags=...P...T
4	4.038400	MS-NLB-PhysServer-10_	Tp-LinkT_2a:84:4e	802.11	111	QoS Data, SN=345, FN=0, Flags=.p.....T
5	4.039428	Tp-LinkT_2a:84:4e	MS-NLB-PhysServer-10_a	802.11	139	QoS Data, SN=198, FN=0, Flags=.p....F.
6	4.141316	MS-NLB-PhysServer-10_	Tp-LinkT_2a:84:4e	802.11	26	QoS Null function (No data), SN=2642, FN=0, Flags=...P...T
7	5.038400	MS-NLB-PhysServer-10_	Tp-LinkT_2a:84:4e	802.11	111	QoS Data, SN=346, FN=0, Flags=.p.....T
8	5.039430	Tp-LinkT_2a:84:4e	MS-NLB-PhysServer-10_a	802.11	139	QoS Data, SN=199, FN=0, Flags=.p....F.
9	5.141316	MS-NLB-PhysServer-10_	Tp-LinkT_2a:84:4e	802.11	26	QoS Null function (No data), SN=2643, FN=0, Flags=...P...T
10	6.039426	MS-NLB-PhysServer-10_	Tp-LinkT_2a:84:4e	802.11	111	QoS Data, SN=347, FN=0, Flags=.p.....T
11	6.040452	Tp-LinkT_2a:84:4e	MS-NLB-PhysServer-10_a	802.11	139	QoS Data, SN=200, FN=0, Flags=.p....F.
12	6.142340	MS-NLB-PhysServer-10_	Tp-LinkT_2a:84:4e	802.11	26	QoS Null function (No data), SN=2644, FN=0, Flags=...P...T
13	8.039426	MS-NLB-PhysServer-10_	Tp-LinkT_2a:84:4e	802.11	111	QoS Data, SN=348, FN=0, Flags=.p.....T
14	8.040964	Tp-LinkT_2a:84:4e	MS-NLB-PhysServer-10_a	802.11	139	QoS Data, SN=201, FN=0, Flags=.p....F.
15	8.143876	MS-NLB-PhysServer-10_	Tp-LinkT_2a:84:4e	802.11	26	QoS Null function (No data), SN=2645, FN=0, Flags=...P...T
16	12.042496	MS-NLB-PhysServer-10_	Tp-LinkT_2a:84:4e	802.11	111	QoS Data, SN=349, FN=0, Flags=.p.....T

WLAN traffic before decryption

In order to decrypt the preceding listed packets, we need to configure the IEEE 802.11 section as follows:

1. Go to **Edit** | **Preferences**, expand the **Protocol** section, select **IEEE 802.11** and configure it as follows:

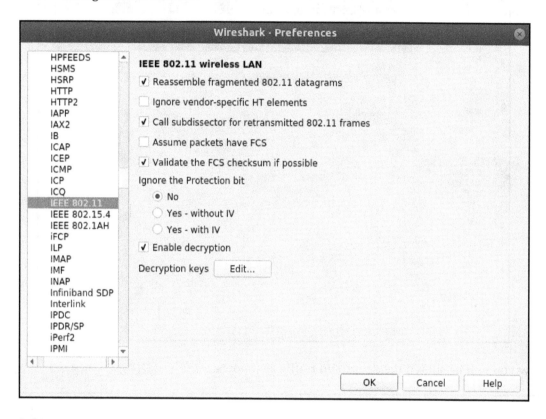

2. Click on the **Edit** button next to **Decryption Keys**.

3. Click on **New** and add the **WEP/WPA** key to enable decryption. After all the changes have been made, click on **OK**:

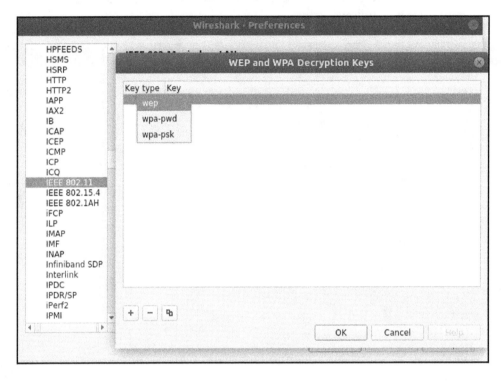

Now you will be shown the decrypted traffic as follows:

No.	Time	Source	Destination	Protocol	Length	Info
2	0.000004	192.168.0.1	192.168.0.100	ICMP	145	Destination unreachable (Network unreachable)
3	0.101892	MS-NLB-PhysServer-10_Tp-LinkT_2a:84:4e		802.11	26	QoS Null function (No data), SN=2641, FN=0, Flags=...P...T
4	4.038400	192.168.0.100	192.168.0.1	DNS	111	Standard query 0xeed6 A ctldl.windowsupdate.com
5	4.039428	192.168.0.1	192.168.0.100	ICMP	139	Destination unreachable (Network unreachable)
6	4.141316	MS-NLB-PhysServer-10_Tp-LinkT_2a:84:4e		802.11	26	QoS Null function (No data), SN=2642, FN=0, Flags=...P...T
7	5.038400	192.168.0.100	192.168.0.1	DNS	111	Standard query 0xeed6 A ctldl.windowsupdate.com
8	5.039430	192.168.0.1	192.168.0.100	ICMP	139	Destination unreachable (Network unreachable)
9	5.141316	MS-NLB-PhysServer-10_Tp-LinkT_2a:84:4e		802.11	26	QoS Null function (No data), SN=2643, FN=0, Flags=...P...T
10	6.039426	192.168.0.100	192.168.0.1	DNS	111	Standard query 0xeed6 A ctldl.windowsupdate.com
11	6.040452	192.168.0.1	192.168.0.100	ICMP	139	Destination unreachable (Network unreachable)
12	6.142340	MS-NLB-PhysServer-10_Tp-LinkT_2a:84:4e		802.11	26	QoS Null function (No data), SN=2644, FN=0, Flags=...P...T
13	8.039426	192.168.0.100	192.168.0.1	DNS	111	Standard query 0xeed6 A ctldl.windowsupdate.com
14	8.040964	192.168.0.1	192.168.0.100	ICMP	139	Destination unreachable (Network unreachable)
15	8.143876	MS-NLB-PhysServer-10_Tp-LinkT_2a:84:4e		802.11	26	QoS Null function (No data), SN=2645, FN=0, Flags=...P...T
16	12.042496	192.168.0.100	192.168.0.1	DNS	111	Standard query 0xeed6 A ctldl.windowsupdate.com

WLAN traffic after decryption

Summary

The IEEE 802.11 standard works over radio frequencies for communication purposes. CSMA/CD facilitates the collision-free environment required for a high-performance wireless networks.

There are commonly three types of frames observed while doing wireless traffic analysis: management, control, and data frames. Management frames control the establishment of the connection, control frames manage the transmission of packets, and data frames consist of the actual data.

Enterprise authentication protocol (**EAP**) in LAN becomes EAPOL, which is used in 802.11 infrastructures (RADIUS/ AAA) for the exchange of master keys.

EAP is used to let the exchange of master keys take place. As defined in RFC 3748, EAP is an authentication framework that supports multiple kinds of authentication methods, and to execute EAP, you do not require an IP because it runs over a data-link layer.

Access points broadcast beacon frames that wireless clients listen for. Also, wireless clients may send a probe request to get connected to the access point, followed by authentication carried out by the access point or third-party authentication service.

Using Wireshark, it is possible to decrypt wireless communications by adding wireless network keys within IEEE 802.11 protocol section.

8
Mastering the Advanced Features of Wireshark

In this chapter, we will look under the hood of the advanced options available in Wireshark and work with a command-line version of packet sniffer. Here, we will be covering the following topics:

- Analyzing the network using the Statistics menu
- Using TCP Stream
- Using the Protocol Hierarchy Option
- Using command-line tools for protocol analysis

With Wireshark, a variety of statistics about the network packets, protocols and endpoints can be viewed and analyzed. Understanding and awareness of advanced features such as protocol hierarchy, conversations, endpoints, and so on, assists in performing tasks pertaining to troubleshooting, optimizing, and forensics activity through viewing and analyzing network related information specifics in detail.

The Statistics menu

Wireshark provides various tools that assist in collecting network stats, which help users in analyzing information ranging from general information to specific protocol-related information.

Using the Statistics menu

Details with respect to the packets captured, filters applied, marked packets, and various other stats can be checked in the Statistics menu; refer to the following screenshot for reference (source: `http://wireshark.org`):

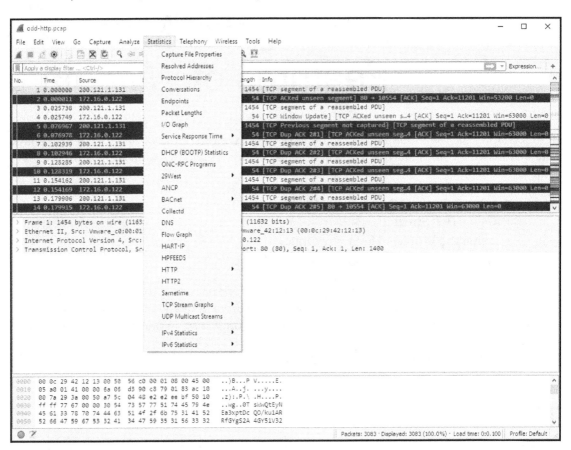

Protocol Hierarchy

The **Protocol Hierarchy** window provides details pertaining to the distribution of protocols seen in network traffic. Each of the rows represents stats pertaining to one protocol; refer to the following screenshot:

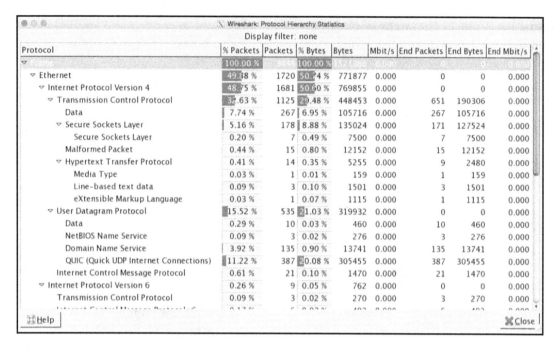

Protocol Hierarchy window

If you want to check the protocol distribution for a specific host, then before you open the **Protocol Hierarchy** window, apply a **Display filter**, for example, `ip.addr==172.20.10.1`. Now, when you open the hierarchy window again the filter will be visible at the top of the **Protocol Hierarchy** window just below the title bar:

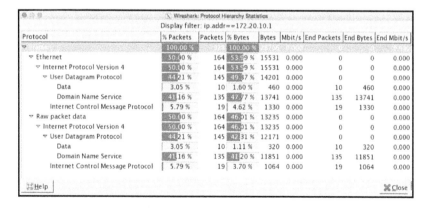

Protocol Hierarchy window after applying display filter

Using the **Protocol Hierarchy** window, display filters can be generated and applied too. Just right-click on the protocol you wish to use and then choose the desired option, as shown in the following screenshot:

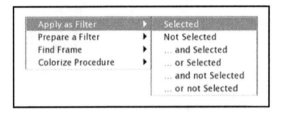

The **Protocol Hierarchy** window will be worth checking in an event where the malware-related activity needs to be assessed and analyzed.

Conversations

To analyze network communication pertaining to two specific endpoints, Conversation option can be used (available under **Statistics** menu). To access it, click on **Statistics** | **Conversations**. The window will list the network layers to assess at the top, and endpoint addresses (IP or MAC) in rows:

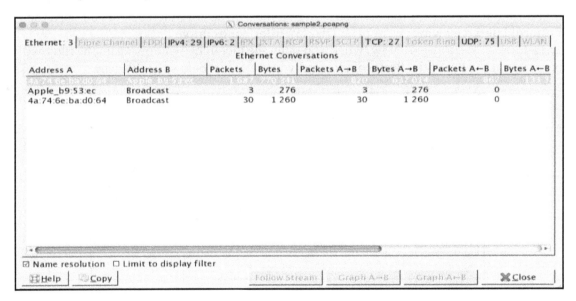

Conversations window

For instance, if we need to identify the endpoint which is generating the most traffic in the network, go to the **IPv4** tab and sort the Bytes column in descending order:

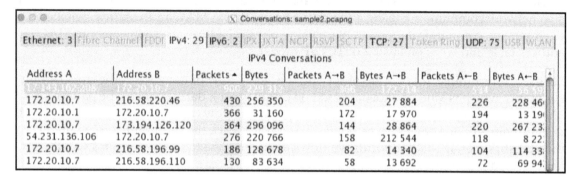

Busiest devices

In the preceding screenshot, the first row depicts how many packets and bytes have been sent and received by the endpoints. For creating a display filter through conversation dialog, right-click on a row and then create the desired expression. I chose the first option, **A<->B**, which would only display packets associated with Address **A** and Address **B**:

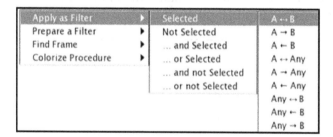

The newly created filter expression will be shown in the **Display Filter** dialog, as shown in the following screenshot:

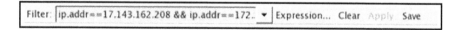

The **Conversations** dialog assists in collecting and analyzing details in the granular form associated with specific endpoints, which comes in handy while troubleshooting and auditing networking infrastructures.

Endpoints

Devices that communicate over a network are referred to as endpoints. Endpoints in a local area network communicate using a physical address that is MAC address. In a switched environment, communication takes place using physical addresses; switches store MAC address table and work on layer 2 of TCP/IP model.

Let's say, for example, that we are observing the heavy flow of network traffic from certain endpoints, which is kind of unusual based on our playbook data (usual traffic pattern). To identify the exact endpoint from which the superfluous flow of network traffic is generated, the **Endpoints** dialog comes to the rescue. To access it, click the **Endpoints** option under the **Statistics menu**. The **Endpoints** windows look quite like the Conversations windows we observed previously.

By default, the **Ethernet** tab will be shown (which lists the layer-2 MAC address) in most cases. Along with the protocol, you must observe a number that states the number of endpoints captured for that specific protocol. In our case, we are seeing **3**, and the same number of rows are visible in the **Main** pane.

In the **Main** pane, many more specific details can be seen for every endpoint, such as the total number of packets transferred, total number of bytes transferred, and total bytes and packets received and transmitted for an individual endpoint:

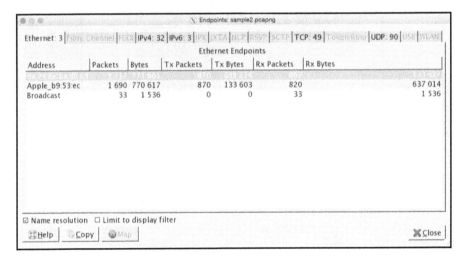

Endpoints window

Now, if you want to analyze other protocols, then simply click on any tab of your choice. I clicked on the **IPv4** tab and sorted the main pane using the **Packets** column, as shown in the following screenshot.

By just looking at the **Endpoints** dialog, I can now easily figure out that the most data was transferred from IP `172.20.10.7`. This could be one single IP talking to some server or, more likely, a server talking to multiple machines on our network at a moderate rate:

Address	Packets ▲	Bytes	Tx Packets	Tx Bytes	Rx Packets	Rx Bytes	Latitude	Longitude
172.20.10.7	3 404	1 518 822	1 752	255 718	1 652	1 263 104	–	–
17.143.162.208	900	229 312	366	172 714	534	56 598	–	–
216.58.220.46	430	256 350	226	228 466	204	27 884	–	–
172.20.10.1	366	31 160	172	17 970	194	13 190	–	–
173.194.126.120	364	296 096	220	267 232	144	28 864	–	–
54.231.136.106	276	220 766	158	212 544	118	8 222	–	–
216.58.196.99	186	128 678	104	114 338	82	14 340	–	–
216.58.196.110	130	83 634	72	69 942	58	13 692	–	–
17.178.104.39	114	45 990	52	29 624	62	16 366	–	–
216.58.196.97	104	34 162	44	19 058	60	15 104	–	–
17.151.236.24	90	28 432	40	20 386	50	8 046	–	–
216.58.196.109	80	35 144	36	17 770	44	17 374	–	–
216.58.196.98	72	28 854	32	16 536	40	12 318	–	–
17.167.194.236	60	14 250	28	10 820	32	3 430	–	–

Ethernet: 3 | Fibre Channel | FDDI | IPv4: 32 | IPv6: 3 | IPX | JXTA | NCP | RSVP | SCTP | TCP: 49 | Token Ring | UDP: 90 | USB | WLAN

Endpoints: sample2.pcapng

IPv4 Endpoints

☑ Name resolution ☐ Limit to display filter

Help · Copy · Map · Close

Endpoints dialog—IPv4 tab

To create a display filter through the Endpoints window, right-click on the row with the most packets transferred and choose **Selected** under **Apply as Filter**, as shown in the following screenshot:

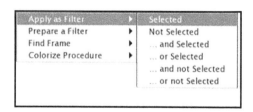

You see a display filter for the same in the **Display Filter** dialog above the **List** pane, like the one shown here:

This facilitates us to quickly analyze traffic for a certain endpoint and hence increases the speed of analysis for users. Once you click on **Clear**, you will be presented with the same **Endpoints** dialog. At the bottom of the window, you will see two checkboxes and a few buttons. The purpose of each is listed below:

- **Name Resolution**: Resolves the name of each of the Ethernet addresses listed in the **Ethernet** tab. But in some scenarios, it might affect the performance of the application adversely, for example, when trying to resolve the unique IP addresses from a huge capture file.
- **Limit to display filter**: Limits the results of the Endpoint window on the basis of a display filter that is applied through the Wireshark main window.
- **Copy**: Copies the content of the current Endpoints window tab in a CSV format (comma-separated values).
- **Map**: Maps the selected endpoint's geographical location in your browser.

Follow TCP Streams

Wireshark provides the feature of reassembling a stream of plain text protocol packets into a human-readable format:

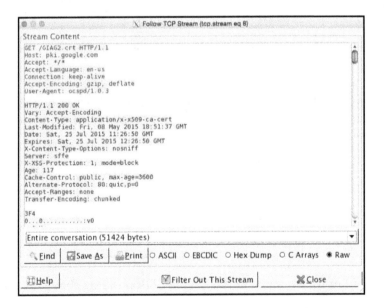

Follow TCP Stream window

For instance, assembling an HTTP session will display the GET requests sent from the client and the responses received from the server. There is specific color coding that is followed by the request and response messages shown in the Follow TCP Stream dialog. Client requests are shown in red, and any text in blue denotes the response received from the server. If the protocol is HTTP, FTP, Telnet, and so on, then the conversation will be shown in plain text; if a secure version of the application layer protocol is used, then some content of the request and response messages will be encrypted.

At the bottom of the Follow TCP stream dialog, a drop-down menu is present from where content in the Follow TCP stream window can be filtered to view only content pertaining to either side of the communication. Also, instead of just viewing the data in RAW format, you can choose between ASCII, EBCDIC, Hex dump, and C arrays format, as desired.

To view the TCP stream, follow these steps:

1. Open the `capture/trace` file
2. Apply the **Display filter** if required
3. Select any packet from the **List** pane
4. Right-click on the selected packet and click on **Follow TCP Stream**

Command line-fu

With the default installation of Wireshark, a command-line version of protocol analyser called Tshark also gets installed. There are a good number of CUI-based sniffing tools available, including Capinfos, Dumpcap, Editcap, Mergecap, Rawshark, Reordercap, Text2pcap, and Tshark.

The most common and widely used command-line tool for protocol analysis purposes is Tshark, which can capture live traffic and analyze saved capture files. Tshark uses the `pcap` library to capture and translate the packets. Just like Wireshark's filtering option are available in Tshark too. Applications like Tshark prove themselves worthy, with benefits such as low memory requirement, easy installation, and simple command sets to run the sniffer.

Let's consider a scenario to understand the usage and advantages of command-line sniffers. Say, for instance, we have an Apache web server and an FTP server running on a Windows box located at IP `172.16.136.128`, and a Macintosh client running at `172.16.136.1`:

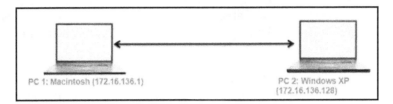

We will start with the basics and eventually move toward the usage of advanced features such as filters and usage of a few of the available statistics options.

Let's try the tool with usage of different features it facilitates:

- The first thing to confirm is how many interfaces are available for capturing packets. Use the following command to check `tshark -D`:

```
Anonymous:Desktop NotFound$ tshark -D
1.  en0 (Ethernet)
2.  fw0 (FireWire)
3.  bridge0 (Thunderbolt Bridge)
4.  utun0
5.  pktap0
6.  en1 (Wi-Fi)
7.  en2 (Thunderbolt 1)
8.  lo0 (Loopback)
```

Interfaces available

- If no interface is specified for capturing network traffic, `tshark` will choose the first interface from the list. Interfaces can be chosen by their names and by the sequence number they appear in.

- For our scenario, we will be using `pktap0` that will listen to the traffic between the client and the server. The command to initiate the capture process is `tshark -i pktap0`:

```
Anonymous:Desktop NotFound$ tshark -i pktap0
Capturing on 'pktap0'
```

- In order to generate some traffic between the client and the server, I have executed the command-line utility `curl` from the client to visit the web page at IP `172.16.136.128`:

```
Anonymous:Desktop NotFound$ curl http://172.16.136.128
```

- As a result of the preceding command, we will see some activity on the Tshark console:

```
Anonymous:Desktop NotFound$ tshark -i pktap0
Capturing on 'pktap0'
  1   0.000000 172.16.136.1 -> 172.16.136.128 TCP 64 51816-80 [SYN] Seq=0 Win=65535 Len=0 MSS=1460 WS
  2 -745883619.684183 172.16.136.128 -> 172.16.136.1 TCP 64 80-51816 [SYN, ACK] Seq=0 Ack=1 Win=64240
  3 -733373297.062554 172.16.136.1 -> 172.16.136.128 TCP 52 51816-80 [ACK] Seq=1 Ack=1 Win=131744 Len
  4 -1830766245.431098 172.16.136.1 -> 172.16.136.128 HTTP 130 GET / HTTP/1.1
  5 -1830766245.129806 172.16.136.1 -> 172.16.136.128 HTTP 130 [TCP Retransmission] GET / HTTP/1.1
  6 -1664501840.066843 172.16.136.128 -> 172.16.136.1 TCP 52 80-51816 [ACK] Seq=1 Ack=79 Win=64162 Le
  7 -392509417.396438 172.16.136.128 -> 172.16.136.1 TCP 52 [TCP Dup ACK 6#1] 80-51816 [ACK] Seq=1 Ac
  8 -2027256734.439159 172.16.136.128 -> 172.16.136.1 HTTP 345 HTTP/1.1 302 Found
  9 -179068134.420122 172.16.136.1 -> 172.16.136.128 TCP 52 51816-80 [ACK] Seq=79 Ack=294 Win=131456
 10 -2067155579.763355 172.16.136.1 -> 172.16.136.128 TCP 52 51816-80 [FIN, ACK] Seq=79 Ack=294 Win=1
 11 -1830766248.828112 172.16.136.128 -> 172.16.136.1 TCP 52 80-51816 [ACK] Seq=294 Ack=80 Win=64162
 12 -392509283.614170 172.16.136.1 -> 172.16.136.128 TCP 52 [TCP Dup ACK 10#1] 51816-80 [ACK] Seq=80
 13 -1830766248.686849 172.16.136.128 -> 172.16.136.1 TCP 52 80-51816 [FIN, ACK] Seq=294 Ack=80 Win=6
 14 -392569681.317465 172.16.136.1 -> 172.16.136.128 TCP 52 51816-80 [ACK] Seq=80 Ack=295 Win=131456
```

Packets captured at pktap0

If you want to stop the capture process at any point, press *Ctrl + C*.

- If you wish to save captured network packets to a file, specify the `-w` switch, as follows:

```
Anonymous:Desktop NotFound$ tshark -i pktap0 -w http.txt
Capturing on 'pktap0'
11
```

- As a result of the preceding command, the raw network data will be stored in a text file named `http.txt`. Following is the content saved in the text file:

```
Anonymous:Desktop NotFound$ cat http.txt


?M<+????????.Mac OS X 10.10.3, build 14D136 (Darwin 14.3.0)4Dumpcap


D136 (Darwin 14.3.0)``???@@E@f?@@k???????lP??f??????
???x``dA???_@@E@7@?},?????P?l?@3?f?????a??
@@q??????lP??f??@1?
???xT??4??9??E??@@H???????lP??f??@1?h
???xGET / HTTP/1.1
User-Agent: curl/7.37.1
Host: 172.16.136.128
Accept: */*
```

Raw data stored in the text file

- To save the captured data in a readable form, just use the redirection operator `">>"` to a file:

```
Anonymous:Desktop NotFound$ tshark -i pktap0
Capturing on 'pktap0'
    1   0.000000 172.16.136.1 -> 172.16.136.128 TCP 64 51816→80 [SYN] Seq=0 Win=65535 Len=0 MSS=1460 WS
    2 -745883619.604183 172.16.136.128 -> 172.16.136.1 TCP 64 80→51816 [SYN, ACK] Seq=0 Ack=1 Win=64240
    3 -733373297.062554 172.16.136.1 -> 172.16.136.128 TCP 52 51816→80 [ACK] Seq=1 Ack=1 Win=131744 Len
    4 -1830766245.431098 172.16.136.1 -> 172.16.136.128 HTTP 130 GET / HTTP/1.1
    5 -1830766245.129806 172.16.136.1 -> 172.16.136.128 HTTP 130 [TCP Retransmission] GET / HTTP/1.1
    6 -1664501840.066843 172.16.136.128 -> 172.16.136.1 TCP 52 80→51816 [ACK] Seq=1 Ack=79 Win=64162 Le
    7 -392509417.396438 172.16.136.128 -> 172.16.136.1 TCP 52 [TCP Dup ACK 6#1] 80→51816 [ACK] Seq=1 Ac
    8 -2027256734.439159 172.16.136.128 -> 172.16.136.1 HTTP 345 HTTP/1.1 302 Found
    9 -179068134.420122 172.16.136.1 -> 172.16.136.128 TCP 52 51816→80 [ACK] Seq=79 Ack=294 Win=131456
   10 -2067155579.763355 172.16.136.1 -> 172.16.136.128 TCP 52 51816→80 [FIN, ACK] Seq=79 Ack=294 Win=1
   11 -1830766248.828112 172.16.136.128 -> 172.16.136.1 TCP 52 80→51816 [ACK] Seq=294 Ack=80 Win=64162
   12 -392509283.614170 172.16.136.1 -> 172.16.136.128 TCP 52 [TCP Dup ACK 10#1] 51816→80 [ACK] Seq=80
   13 -1830766248.686849 172.16.136.128 -> 172.16.136.1 TCP 52 80→51816 [FIN, ACK] Seq=294 Ack=80 Win=6
   14 -392569681.317465 172.16.136.1 -> 172.16.136.128 TCP 52 51816→80 [ACK] Seq=80 Ack=295 Win=131456
```

As a result of issuing the preceding command, packets are captured and redirected to the text file `http2.txt`. Following is the content saved in the text file, that lists the packets captured between the two hosts `172.16.136.128` and `172.16.136.1` over port `80`:

```
Anonymous:Desktop NotFound$ cat http2.txt
  1    0.000000 172.16.136.1 -> 172.16.136.128 TCP 64 51821-80 [SYN] Seq=0 Win=65535 Len=0 MSS=1460 WS=32
  2  -1830767469.040043 172.16.136.128 -> 172.16.136.1 TCP 64 80-51821 [SYN, ACK] Seq=0 Ack=1 Win=64240 L
  3  -1830767469.040009 172.16.136.1 -> 172.16.136.128 TCP 52 51821-80 [ACK] Seq=1 Ack=1 Win=131744 Len=0
  4  -2016764535.847514 172.16.136.1 -> 172.16.136.128 HTTP 130 GET / HTTP/1.1
  5  -2027256734.427691 172.16.136.128 -> 172.16.136.1 HTTP 345 HTTP/1.1 302 Found
  6  -1830767469.037172 172.16.136.1 -> 172.16.136.128 TCP 52 51821-80 [ACK] Seq=79 Ack=294 Win=131456 Le
  7  -1830767469.037084 172.16.136.1 -> 172.16.136.128 TCP 52 51821-80 [FIN, ACK] Seq=79 Ack=294 Win=1314
  8  -1935145592.773838 172.16.136.128 -> 172.16.136.1 TCP 52 80-51821 [ACK] Seq=294 Ack=80 Win=64162 Len
  9  -1830767469.036949 172.16.136.1 -> 172.16.136.128 TCP 52 [TCP Dup ACK 7#1] 51821-80 [ACK] Seq=80 Ack
 10  -1935145592.773838 172.16.136.128 -> 172.16.136.1 TCP 52 80-51821 [FIN, ACK] Seq=294 Ack=80 Win=6416
 11  -1830767469.036570 172.16.136.1 -> 172.16.136.128 TCP 52 51821-80 [ACK] Seq=80 Ack=295 Win=131456 Le
```

We just learnt the two different ways to save network packets to a file.

- Tshark facilitates three types of filtering options: Capture, Display, and Read. We have discussed the Capture and Display filters in earlier chapters, so let's discuss the read filter. The read filter is able to filter traffic from live as well as save captured files. Through read filters a particular set of packets can be decoded or written to a file.
- Using the Read filter is a processor-intensive task, and issues like packet loss could be observed, and capture filters are preferred over read filters. For the capture filter the `-f` switch is used; `-R` is used for the read filter; and `-Y` is used for the display filter. Let's learn the usage of the capture filter using `-f` switch:

Usage of a switch is case-sensitive.

```
Anonymous:Desktop NotFound$ tshark -i pktap0 -f "port 20"
Capturing on 'pktap0'
  1    0.000000 172.16.136.1 -> 172.16.136.128 TCP 64 51852-20 [SYN] Seq=0 Wi
  2    0.000151 172.16.136.128 -> 172.16.136.1 TCP 64 20-51852 [SYN, ACK] Seq
  3  -1438261061.117554 172.16.136.1 -> 172.16.136.128 TCP 52 51852-20 [ACK]
  4  -565845755.905104 172.16.136.128 -> 172.16.136.1 FTP-DATA 94 FTP Data: 4
  5    0.330476 172.16.136.1 -> 172.16.136.128 TCP 52 51852-20 [ACK] Seq=1 Ac
  6  -1438260168.702253 172.16.136.128 -> 172.16.136.1 FTP-DATA 97 FTP Data:
  7  -776735948.749363 172.16.136.1 -> 172.16.136.128 TCP 52 51852-20 [ACK] S
```

- Use double quotes around the filter expression, if the desired expression has space character like shown in preceding screenshot for example `"port<space>20"`.
- Now, let's learn the usage of the display filter over a previously saved capture file `http.pcap`, and filter all HTTP packets originating from the web server at IP `172.16.136.128`:

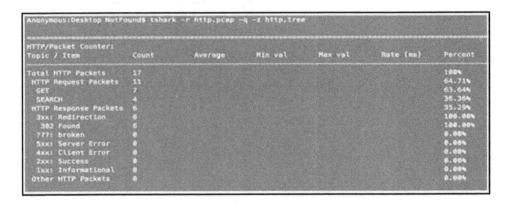

Tshark display filter

- In order to collect the HTTP protocol, only statistics from the `http.pcap` file use the command `tshark -r <file-name> -q -z <expression>`:

```
Anonymous:Desktop NotFound$ tshark -r http.pcap -q -z http,tree

HTTP/Packet Counter:
Topic / Item           Count    Average    Min val    Max val    Rate (ms)    Percent

Total HTTP Packets     17                                                     100%
 HTTP Request Packets  11                                                     64.71%
  GET                  7                                                      63.64%
  SEARCH               4                                                      36.36%
 HTTP Response Packets 6                                                      35.29%
  3xx: Redirection     6                                                      100.00%
   302 Found           6                                                      100.00%
  ???: broken          0                                                      0.00%
  5xx: Server Error    0                                                      0.00%
  4xx: Client Error    0                                                      0.00%
  2xx: Success         0                                                      0.00%
  1xx: Informational   0                                                      0.00%
 Other HTTP Packets    0                                                      0.00%
```

- The `-q` switch keeps it silent over the standard output (this is generally used while working with statistics in Wireshark) and the `-z` switch is used for activating various statistics options. Both switches are often used in conjunction.

- If you want to check how many hosts were observed while capturing the network traffic, use the following command:

```
Anonymous:Desktop NotFound$ tshark -r http.pcap -q -z hosts
# TShark hosts output
#
# Host data gathered from http.pcap

172.16.158.1    Anonymous.local
172.16.136.1    Anonymous.local
```

Tshark is a powerful yet simple command-line sniffer which is similar to `tcpdump`. It enables capturing of network packets with ease and less configuration/installation required.

Summary

The Conversations window lists information pertaining to communication between two hosts.

The Endpoints dialog lists details pertaining to the devices connected to the network.

Wireshark Summary is an informational feature, which offers a granular form of data, filters, and the `trace` file.

The Protocol Hierarchy window lists information in a tabular format pertaining to distribution of protocols used by the network endpoints.

Use the Follow TCP Stream option in Wireshark to read the plain text data from captured packets. There are different viewing options available such as ASCII, and Hex.

A command-line tool gets installed when you install Wireshark. The most common tool used is Tshark, which works in a similar way to Wireshark and `tcpdump`. It uses the pcap library that is used by other major protocol analyzers.

With Tshark, you can listen to live networks or work with an already saved capture file.

Other Books You May Enjoy

If you enjoyed this book, you may be interested in these other books by Packt:

Mastering Wireshark 2
Andrew Crouthamel

ISBN: 978-1-78862-652-1

- Understand what network and protocol analysis is and how it can help you
- Use Wireshark to capture packets in your network
- Filter captured traffic to only show what you need
- Explore useful statistic displays to make it easier to diagnose issues
- Customize Wireshark to your own specifications
- Analyze common network and network application protocols

Network Analysis using Wireshark 2 Cookbook - Second Edition
Nagendra Kumar Nainar, Yogesh Ramdoss, Yoram Orzach

ISBN: 978-1-78646-167-4

- Configure Wireshark 2 for effective network analysis and troubleshooting
- Set up various display and capture filters
- Understand networking layers, including IPv4 and IPv6 analysis
- Explore performance issues in TCP/IP
- Get to know about Wi-Fi testing and how to resolve problems related to wireless LANs
- Get information about network phenomena, events, and errors
- Locate faults in detecting security failures and breaches in networks

Leave a review - let other readers know what you think

Please share your thoughts on this book with others by leaving a review on the site that you bought it from. If you purchased the book from Amazon, please leave us an honest review on this book's Amazon page. This is vital so that other potential readers can see and use your unbiased opinion to make purchasing decisions, we can understand what our customers think about our products, and our authors can see your feedback on the title that they have worked with Packt to create. It will only take a few minutes of your time, but is valuable to other potential customers, our authors, and Packt. Thank you!

Index